CAMARO CLASSICS

CAMARO CLASSICS

BY THE AUTO EDITORS OF CONSUMER GUIDE®

Publications International, Ltd.

Louis Weber, CEO
Publications International, Ltd.
7373 North Cicero Avenue
Lincolnwood, Illinois 60712

ISBN-13: 978-1-4508-2433-0
ISBN-10: 1-4508-2433-1

Manufactured in China.

8 7 6 5 4 3 2 1

Library of Congress Control Number: 2011922736

Credits

Photography:
The editors would like to thank the following people and organizations for supplying the photography that made this book possible. They are listed below, along with the page number(s) of their photos.

Jeff Cohn: 35, 125; **Roland Flessner:** 91; **Mitch Frumkin:** 105, 107, 109, 115; **Sam Griffith:** 15, 29, 37, 117; **Brandon Hemphill:** 43; **Bud Juneau:** 21, 47, 51; **Milton Kieft:** 27; **Nick Komic:** 13, 19; **Dan Lyons:** Back cover, 83, 113, 121; **Vince Manocchi:** 25, 61, 89, 119, 128; **Doug Mitchel:** 17, 39, 45, 49, 53, 59, 65, 67, 69, 71, 73, 77, 123; **Mike Mueller:** 57, 81; **Jeff Rose:** 23; **David Temple:** 63, 75, 103; **Phil Toy:** 33; **W. C. Waymack:** 31, 41, 87; **Nicky Wright:** Back cover, 11, 55, 101

Owners:
Special thanks to the owners of the cars featured in this book for their cooperation. Their names and the page numbers for their vehicles follow:

Tom Bigelow: 11; **Jerry and Carol Buczhowski:** Back cover, 55, 101; **Chicago Car Exchange:** 45; **Mike Coffee:** Back cover, 83; **Donald R. Crile:** 27; **Dr. Mike Cruz:** 61; **Ed Cunneen:** 37; **Mike D'Amico:** 39; **Sandy D'Amico:** 49; **Tony D'Amico:** 53; **Mitch Deck:** 73; **DeNooyer Chevrolet:** 121; **Rob Embleton:** 15, 29; **Ernest Gigliotti:** 9; **Robert Harand:** 25; **Debra Hogan:** 81; **Justin Hurn:** 41; **Mark Knecht:** 35; **Steven Knutsen:** 51; **Bryan Leinart:** 17; **Bob Lichty/Motorcar Portfolio:** 43; **Charley Lillard:** 47; **Ken McMullen:** 31; **Ann M. Mocklin:** 77; **Jason Newman:** 103; **John R. Oehler:** 65, 67, 69, 71; **D. R. Ogsberger:** 23; **Tom Pellett:** 63; **Ramshead Automobile Collection:** 21; **Ron and Laureen Roach:** 33; **Steve Schultz/SS Classic Cars:** 13, 19; **Ralph Segars:** 75; **Jeff Sweet:** 113; **Wayne Thuenemann:** 87; **Volo Auto Museum:** 59; **Barry Waddell:** 57

Our appreciation to the historical archives and media services of General Motors Company.

CONTENTS

Foreword 6
1967 RS/SS Convertible 8
1967 RS Coupe 10
1967 Yenko Super Camaro 12
1967 I-6 Convertible 14
1968 Coupe 16
1968 Yenko Super Camaro SS 18
1968 SS Coupe 20
1968 Z/28 RS Coupe 22
1968 Coupe 24
1968 Z/28 RS Coupe 26
1969 I-6 Base Coupe 28
1969 I-6 Convertible 30
1969 SS Coupe 32
1969 Z28 34
1969 ZL-1 36
1969 RS Coupe 38
1969 RS Convertible 40
1969 327 Convertible 42
1969 396 SS Coupe 44
1969 396 SS Convertible 46
1969 Indy Pace Car Convertible Replica 48
1969 SS Z10 Coupe 50
1969 RS/Z28 52
1969 Yenko SC Coupe 54
1970 Z28 Coupe 56
1970 Z28 Coupe 58
1970 Z28 Hurst "Sunshine Special" Sunroof Coupe 60
1970 SS/RS Coupe 62
1971 Z28 RS 64
1974 Z28 Coupe 66
1977 Z28 Coupe 68
1978 Z28 Coupe 70
1979 Z28 Coupe 72
1979 Berlinetta Coupe 74
1980 Z28 Coupe 76
1981 Yenko Turbo Z 78
1981 Berlinetta Coupe 80
1982 Indy Pace Car Replica 82

1983 Z28 Coupe 84
1984 Z28.................................... 86
1986 IROC-Z................................. 88
1986 IROC-Z T-top 90
1987 Z28 Coupe 92
1987 Z28 Convertible......................... 94
1987 IROC-Z................................. 96
1988 IROC-Z................................. 98
1990 IROC-Z Coupe.......................... 100
1991 RS V-6 Coupe 102
1991 Z28 Coupe 104
1992 25th Anniversary Z28 106
1992 25th Anniversary RS Convertible 108
1993 Z28 Coupe 110
1993 Z28 Indy Pace Car Replica 112
1994 Z28.................................... 114
1995 Z28 Convertible........................ 116
1997 30th Anniversary Z28 Coupe 118
1998 Coupe.................................. 120
2002 35th Anniversary SS 122
2010 Coupe.................................. 124
2011 Convertible............................ 126

FOREWORD

It's an oversimplification to say that Chevy's Camaro was GM's "answer" to the Ford Mustang. Granted, Camaro's September 1966 appearance in showrooms trailed Mustang's debut by 29 months—and yet Camaro evolved largely independently of influence from Ford's pony car.

In fact, the roots of Camaro are in the Buick Riviera, a midsize "personal car" introduced for the 1963 model year. The Riv appealed to Chevy design chief Irv Rybicki because when he looked at it he saw the potential for another, smaller car that could be profitably badged as a Chevrolet.

Production of an existing small Chevy, the faux-sporty Corvair Monza, had expanded significantly because of heavy promotion during 1960-61. However, the sales boost wasn't attributable in any appreciable way to the growing youth market. Monza wasn't quite what young drivers were looking for.

GM's overall chief of design, Bill Mitchell, agreed with Rybicki's assessment that another car was hidden in the Riv, and so, in 1962, Rybicki's design staff began secret work on what was initially code-named "Panther." In much the way that Mustang evolved from the Falcon platform, early Camaro development was based on the compact Chevy II.

The Panther team came up with a long-hood/short-deck design very much like Mustang's—a good trick because no one on the Panther project knew anything about a proposed production car called Mustang.

The 1967 Camaro hit the sweet spots of performance, style, and youthful appeal. It was a strong seller from the outset and remained a potent rival to Mustang—in the marketplace and on track and strip—for decades.

When coupe sales declined after the millennium, Camaro was discontinued, but it returned, with eye-opening panache and strong sales, as an exciting 2010 model. America still wants to hug the Hugger.

SS
SS
CAMARO

Command Performance

You don't drive it; you command it!

Chevrolet's new Camaro is here! It's Corvette excitement with back-seat passenger room, a sporting hustler that's nothing but lean and forceful.

Camaro is your kind of car. It has wide-stance design for flattening corners, slim Strato-bucket seats that hang on comfortably, your choice of convertible or sport coupe and a big-car engine lineup to quicken your pulse. It starts at a base six with 140 horsepower and goes up through three V8s including an exclusive new 350-cubic-inch power plant.

After that, Camaro is up to you. Choose from a series of packages you can add to make it exactly your kind of car.

Go the Rally Sport route and dramatically change the appearance of Camaro. Make it a no-nonsense sporting machine with the SS 350 package. Or add both of them and get the best of both worlds.

Then go have one great time!

Top left: Camaro Rally Sport Coupe. Top right: Camaro Sport Coupe. Bottom: Camaro SS 350 Convertible with Rally Sport equipment.
All with seat belts front and rear (retractors on front) and shoulder belt anchors.

How commanding do you want to be?

Camaro

- Twin slimline Strato-bucket seats
- Businesslike round white-on-black instruments
- Single-unit headlights in a black grille
- Long, low hood line and a tucked-in tail
- Choice of 15 Magic-Mirror acrylic colors
- Choice of four color-keyed vinyl interiors, all fully carpeted
- Safety features like dual master cylinder brake system and GM-developed energy-absorbing steering column

Camaro Rally Sport

- Electrically operated hideaway headlights in a full-width black grille
- Wider body sill molding to accent the sporting lines
- Special identification emblems on grille, fenders and filler cap
- Bright metal moldings on roof gutter and wheel openings
- Special taillight treatment
- Color-keyed body side accent stripes

Camaro SS 350

- Exclusive 295-horsepower 350-cubic-inch V8
- Special hood with raised louvers
- Unique striping around grille opening
- Red stripe tires on 14" x 6" wheels
- Special SS identification emblems

And you can add

- Special-purpose front and rear suspension
- Fully synchronized 3-speed or 4-speed floor-mounted gearshift or Powerglide automatic with any engine
- Special package of ammeter, temperature, oil pressure and fuel gauges and clock on console, plus tachometer and light for low fuel on instrument panel
- Speed warning unit
- Front disc brakes

1967

RS/SS Convertible

Chevy's first-year Camaro turned heads when it arrived in showrooms on September 21, 1966. Like rival Mustang, it could be tailored to suit the buyer's needs and temperament. Camaro offered a pair of inline sixes (194 and 230 cid) and V-8s in displacements of 302, 327, 350, and 396 cubic inches. During development of the convertible, torsional vibration was solved with "cocktail shakers"—harmonic shock absorbers mounted at each corner. The fix wasn't unique to Camaro, or even to GM: Lincoln and Thunderbird had come up with something similar for their 1961 models, and Chevy's Corvair got the technology beginning in 1965. This RS/SS Camaro convertible that gleams in its original Madeira Maroon paint runs with the optional 295-horsepower 350 and has the characteristic RS blackout grille with hidden headlamps.

Camaro Custom Interior in red.

Custom Interior

HERE ARE ALL THE COMFORT AND SPORT YOU COULD WANT INSIDE, TOO!

Personalizing doesn't stop on the outside. You can further individualize the interior of your Camaro by specifying the Custom Interior. When you ask for this plush assemblage, you'll get your choice of seven all-vinyl interiors (see the color and interiors descriptions on pages 18-19). The thin line Strato-buckets are further enhanced with color-keyed accent bands framing the embossed seat and backrest panels. The rear seat repeats the theme. And in sport coupes, you can specify an all-vinyl Strato-back front seat in three color choices with individual backrests, color-keyed accent bands and folding center armrest.

When you order the Custom Interior you get: door trim panels of molded vinyl with integral armrests, recessed door handles and carpeted scuff panels. Rear armrests with ash-trays in sport coupes along with courtesy lights in the rear quarter roof panels (courtesy lights are mounted under the instrument panel in convertibles). Special oval-styled steering wheel, molded luggage compartment mat and special underhood insulation complete the Custom Interior line-up.

When you specify Rally Sport or SS 350 Camaro, you get "rs" or "SS 350" emblems in the steering wheel center.

Personalizing your Camaro is one of life's more pleasant experiences. Check the list on pages 16-17 for items you can order; many items are also covered in text and illustrations throughout the catalog.

Full-length armrest with recessed door handles and carpeted scuff panel highlight Custom Interior doors.

1967

RS Coupe

Camaro offered about 80 factory options for 1967, with another 40 accessories that were available as dealer add-ons. Chevy's 327 V-8 developed 275 horsepower in top Camaro trim (a 210-hp variant was also available) and was a popular engine choice. Testers from *Road & Track*, running a 327 RS coupe with 4-speed manual, accomplished the zero-to-60-mph run in 9.1 seconds. The familiar "bumblebee" nose stripe was originally part of the SS option package but became more widely available as a stand-alone option in the spring of 1967. Nearly 65,000 RS Camaros were produced for 1967. Note the side vent windows; '67 was the only model year that Camaro had them.

SPAN-INC. Presents the

1967 YENKO SPORTSCARS

Sold Exclusively Through A Nation-Wide Network of YENKO SPORTSCAR DEALERS

YENKO SUPER CAMARO

Engineered by Dick Harrell

YENKO SUPER CAMARO SPECIFICATIONS

YENKO SUPER CAMARO/450
427CID—450 H.P. V-8 Engine.
Mechanical Lifters.
4 speed close ratio transmission.
Positraction rear end 3.73:1 ratio. Metallic Brakes.
Dual exhaust system—with Deep Tone Mufflers.
Special Heavy-Duty clutch and pressure plate.
Special purpose suspension. Extra capacity radiator.
Chrome valve covers. Full sweep tachometer.
Full instrument package.
14" x 6" Steel wheels.
Wide oval nylon tires.
$4245.00

YENKO SUPER CAMARO/410
427CID—410 H.P. V-8 Engine.
Hydraulic Lifters.
YENKO Special "SUPER GT HYDRO".
3 speed automatic tranmission.
Positraction rear end 3:35 ratio.
Metallic Brakes.
Dual Exhaust system—with Deep Tone Mufflers.
Special purpose suspension. Extra capacity radiator.
Chrome valve covers. Full Sweep Tachometer.
Full instrument package. 14" x 6" Steel wheels.
Wide oval tires. **$4245.00**

Z-28 CAMARO Competition version "THE STORMER"
Full specifications and prices available upon request.
BASE **$6000.00**

SPECIAL OPTIONS AVAILABLE (designed by Don Yenko and Dick Harrell)
SCATTER SHIELD . . . TRACTION MASTERS . . . TUNED EXHAUST . . . L88 ENGINE OPTION . . . SIDE EXHAUST . . . SPECIAL YENKO 2-4 BARREL HIGH RISE INTAKE . . .SPECIAL LIGHTWEIGHT FIBERGLAS HOOD WITH '67 CORVETTE SCOOP . . . FIBERGLAS REAR DECK WITH SPOILER, SPECIAL REAR END RATIOS 4:10, 4:56, 4:88 . . . SPECIAL ALUMINUM OR MAG WHEELS . . . COMPLETE ENGINE AND CHASIS FACILITIES AVAILABLE — ALL COMPETITION CARS TRACK OR STRIP TESTED.

SPAN-INC. Exclusive U.S. Distributor · Sales Office: 300 N. State, Chicago, Ill. 60610

1967

Yenko Super Camaro

Pennsylvania racer and new-car dealer Don Yenko began his performance-mods sideline business with the so-called Yenko Stinger, a ferocious Chevrolet Corvair that saw production of 125 units between 1966 and 1969. In 1967, Yenko and a new partner, conversion specialist Dick Harrell, saw promise in the Camaro and subsequently produced 60 Yenko Super Camaro 450s (so named because the NHRA ran Chevy 427s tweaked to produce 450 hp). Yenko Camaros equipped with automatic transmissions got the 427, with horsepower boosted by Yenko-Harrell from 390 to 410. Super Camaros running with 4-speeds were fitted with an L72 long-block 427, or a 427 short block. Horsepower on each was bumped to 425. Even at the milder 410 horsepower, a Yenko conversion outstripped top factory Camaro hp by 85. Although the Yenko-Harrell Camaros are highly regarded, they weren't the first super-Camaros, as they came after cars modified by Bill Thomas Race Cars, Inc. and Dana Chevrolet, both in California, and from Chicago-based Nickey Chevrolet. In fact, Dick Harrell had a hand in the Dana and Nickey conversions.

427

camaro
427

Try Camaro-"The Hugger"

Camaro hugs a road closer, straightens a curve easier because it's the widest stance sportster at its price. It's lower, heavier, too . . . big-car solid and steady. You get a better ride, more precise handling for your money. Ask any Camaro owner, he'll tell you.

Now, during the Camaro Pacesetter Sale, you also get special savings on specially equipped sport coupes and convertibles. Save on all this: the 250-cu.-in. Six, whitewalls, wheel covers, bumper guards, wheel opening moldings, body striping, deluxe steering wheel, extra brightwork inside. And, at no extra cost during the Sale, get a floor shift for the 3-speed transmission and the sporty hood stripe! Compare Camaro. See your Chevrolet dealer now. (Sale savings, too, on specially equipped Fleetside pickups, Model CS 10934.)

1967

I-6 Convertible

A six-cylinder Camaro was appealing for numerous reasons, not least because a convertible six could be had for $2704—$105 less than for a ragtop with a V-8. Because production of '67 Camaro sixes was modest relative to V-8 production (58,808 to 162,109), and because many surviving examples became "project cars" that were uprated to V-8 power, original inline-six Camaros are rare and desirable today. They look terrific, exhibit pleasing road manners, and are free of the axle tramp that characterizes the first-year big-engine models. This example runs with the 250-cid six, rated at 155 horsepower. The "bumblebee" nose stripe, originally reserved for SS models, became an across-the-board option as model-year production went on.

Chevrolet

DAYTONA
SPORT 60

DAYTONA
SPORT 60

610
SDC 150
DAYTONA
SPORT 60

Outside you'll discover much to please.

A cluster of new and handsome touches are bound to make "the hugger" more appealing than ever in 1968. They add to its beauty, its utility and its distinctiveness. Examine a few of the finer points with us and see if you don't agree that Camaro has outdone itself.

Starting at the front, the Camaro grille sports some modest changes: new satin-silver horizontal bars over deep-set black for a wider look; headlights and new rectangular parking lights that are set into the grille (the Rally Sport version has something special in this department); new, too, are taillights with integral back-up lights. And how about this? . . . ventipanes have given way to single curved side windows for added style and improved visibility. New side marker lights front and rear help stake out your claim to a parking place on darkened streets and show your position more clearly while under way.

It all adds up to a remarkably appealing new wide-stance Camaro for 1968 that won't be content until it entices you inside.

New Camaro grille with inset head and parking lights

New Camaro taillights with integral back-up lights

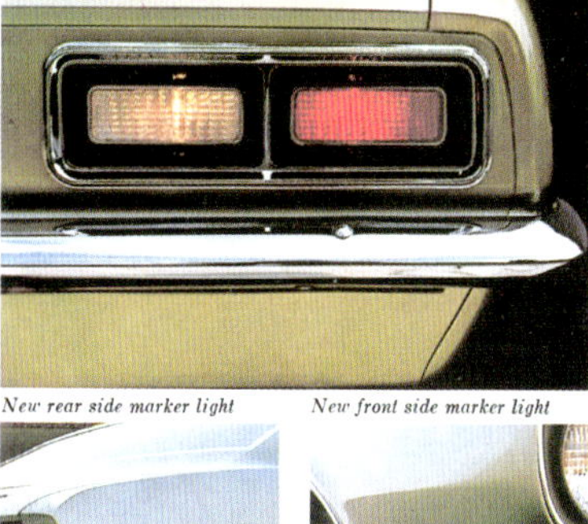

New one-piece curved side window

New rear side marker light *New front side marker light*

The great profile . . . even more finely sculptured for 1968

1968

Coupe

Camaro prices rose by about $100 on all models for 1968, but that didn't deter excited buyers, whose enthusiasm bumped total production to 235,115 units for '68, up from 220,917 for 1967. Camaro lost its vent window this year and picked up federally mandated side-marker lights. More significantly, '68 Camaros had staggered rear shocks that reduced wheel hop—the driver's-side unit was sited ahead of the axle; the passenger-side shock behind it. The restyled grille seen on non-RS models carried a midline vertical crease that subtly divided the grille's contours. At the rear, new center dividers on the taillight pods gave a two-and-two effect, and backup lights (on non-RS models) were incorporated into the larger taillight units. The coupe, opposite, in Matador Red, has the 275-horse 327 V-8.

Camaro

327

DON YENKO
GERRY THOMPSON
DICK THOMPSON
BOB GROSSMAN

"HOME OF THE WINNERS"

DONNA MAE MINNS
ED HEDRICK
PAUL MOYER

SPORTSCARS

454-LS-7—Camaros & Chevelles 350-LT-1 NOVAS—"The Yenko Deuce"

World's Largest Manufacturer of Chevy-Based Hi-Performance Vehicles

Z-28s
SUPER
SPORT
chevelles
novas
camaros
IN STOCK

ALL "HI-PO" SERVICES

COMPLETE LINE OF HI-PO PARTS
Chevrolet and Speed Shops

DYNO-TUNING
Chassis & Engines

RACE CAR PREP
Strip—Road Course—Oval

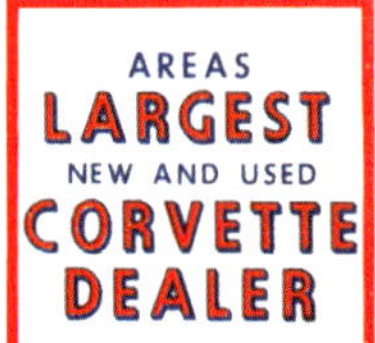

575 WEST PIKE ST., CANONSBURG, PENNSYLVANIA 15317

(412) 745-0100 (412) 745-0100

57-0246-B

1968

Yenko Super Camaro SS

For '68, RS and SS cars got the Yenko treatment. Unfortunately, Don Yenko and engine-swap genius Dick Harrell parted ways at the end of 1967. Because of the split, engines and other components of very nearly all of the '68 conversion Camaros were created by Chevy, for NHRA homologation purposes, and shipped to Yenko for installation. The understated Island Teal example seen here is an SS fitted with Chevy's 427-cid, 425-horse V-8 mated to a Muncie M-21 4-speed transmission. Other goodies include a Warner 970 tach, a 140-mph speedometer, and Pontiac Rally II wheels with "Y" center caps. Total Super Camaro production amounted to 65 for the year, and 40 years later only 14 were known to survive.

YENKO
Camaro
SS

Tough act to follow.

We raised the curtain on Corvette 15 years ago. It has enjoyed the role as America's only true production sports car ever since. Unchallenged. With that lean aerodynamic profile and credits like four-wheel disc brakes, a fully independent suspension system and V8 choices that deliver up to 435 horsepower, it's no wonder. Try one.

Encore!

Then try Camaro. You'll get plenty of Corvette excitement. From the way Camaro looks to its road-hugging ride to its Astro Ventilation to V8s you can order all the way up to 325 horsepower. How do you follow an act like Corvette? Camaro. The Hugger. Produced and directed by Chevrolet.

CHEVROLET

1968

SS Coupe

To distinguish one '68 Camaro SS from another takes a practiced eye: No other Camaro from that year shared a hood with the SS, and SS models had two distinct hoods, one for the SS 350 (a return of what had been the only SS hood for 1967) and another for the SS 396. (For the record, the 350 hood has a pair of decorative ribbed inserts while the 396 bonnet has inserts that are more aggressively styled to mimic the look of carburetion stacks.) A 396 V-8, rated at the standard 325 horsepower, motivates this Butternut Yellow SS coupe with white interior and numerous extra-cost options, including Turbo Hydra-Matic transmission, power steering, power brakes (front discs were now standard with SS), air conditioning with heavy-duty radiator, and a fold-down Strato-back front seat.

Camaro
SS
SPORT G/T

396

1968

Z/28 RS Coupe

The '67 Z/28 was available in white only, but the body-color palette was expanded for 1968, as attested to by this colorful Z/28 with the Rally Sport package and the potent 302 Turbo-Fire V-8 rated at 290 hp. Because of its power-to-weight ratio, the 302 was widely campaigned by Chevrolet and independent race teams and did well against similarly equipped Mustangs and other competitors. Although commonly seen on Z/28s this year, the rear spoiler (which the Chevy options book called "an auxiliary panel & valance") did not come standard but was a $32.65 extra that could be had with any Camaro. For the model year, more than 15,500 of the fiberglass spoilers were sold.

Camaro
BY CHEVROLET

WIDE BOOTS
New Wide Tread tires from Goodyear
You could buy Goodyear's new Wide Boots because their tread is almost one-third wider than the tread on ordinary tires. Or because they start faster. Stop quicker. Handle surer. Corner safer.
You could buy Goodyear's new Wide Boots because they're made much like a racing tire. Squat. Broad shouldered. With a strong cord set at a low angle for less heat buildup and longer wear. With a tread of Tufsyn rubber – the toughest rubber Goodyear ever built into a tire.
Or you could buy Goodyear's Wide Boots—in red or white stripe—just because they look great.
GOODYEAR
See the Winners go Goodyear in
Grand Prix
Presented by MGM

1968

Coupe

The middle of the 1968 model year brought Camaro and other GM cars a new semi-automatic transmission called Torque Drive. It was essentially a low-cost variant of the fully automatic two-speed Powerglide but required manual "shifts." Although some '68 Camaros were fitted with four-wheel discs brakes, the system wasn't a factory option, but something specially installed by dealers or the cars' owners. When Camaro lost its side vent window for 1968, GM brought on AC Delco's new air-conditioning system, which was dubbed "Astro-Ventilation." The '68 coupe seen here is finished in Grotto Blue and has a Medium Blue interior. It runs with the optional 210-horsepower 327-cid V-8 mated to the Powerglide. The coupe left the factory with the Z21 trim package that included bright stainless window surrounds (with drip gutter molding) and bright molding at the beltline and wheel openings.

1968

Z/28 RS Coupe

Z/28 race activity is even more apparent given the history of this LeMans Blue example with the 290-horse 302 mated to a 4-speed stick. Earlier in its life, it had been a racer as well as a show car; it was later restored to stock. This one has the rare woodgrain plastic steering wheel (only 5649 sold) and a clever secondary-gauge package that combined the clock with a tachometer and became known as the "tick-tock-tach." The front-end "bumblebee" band wasn't available with the Z/28 or SS packages, and model-year 1968 was the last to mount the ignition on the dash. The hidden headlight covers were now vacuum-operated, rather than electrical, as they had been for model-year '67.

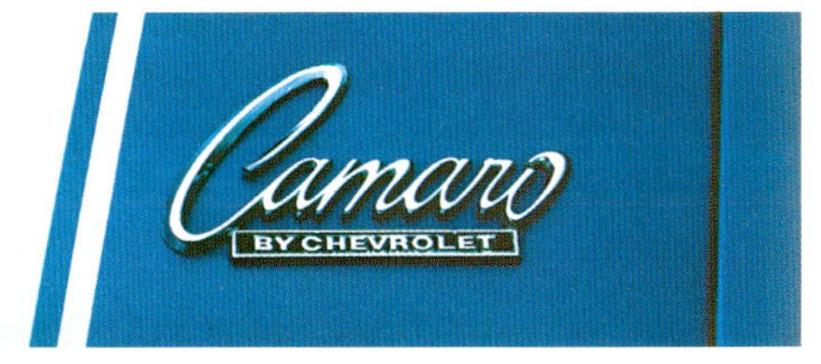
Camaro
BY CHEVROLET

USA-1
What the younger
generation's coming to—
the 1969 CAMARO!
Please call me about
your
Very Important.
George Martens
PUTTING YOU FIRST
KEEPS US FIRST!
LITHO IN U.S.A.

1969

I-6 Base Coupe

Although the final iteration of Camaro's first-generation design, the 1969 was significantly facelifted: at the doors, front fenders, rear quarters, and aft panel. Up front, the grille became more angular, though the general wide-mouth shape was retained. Two-tone paint was new this year; likewise variable-ratio steering that brought more useful movement to the front wheels. Inside, the main instrument panel was redesigned with two primary pods flanking a smaller center one. Instruments on the optional console were unchanged. Chevy sixes were known—and appreciated—for their slow-revving, long-lived natures. The base motor/transmission setup was a 140-hp 250-cid inline six mated to a 3-speed standard manual. Top transmission was a 3-speed Turbo Hydra-Matic. In stripper form, the base coupe could be had for $2621.

The command post.

Camaro, from the rear.

New grille and parking lights.

CAMARO

If Camaro looks longer and wider this year, that's because it is. Result: a steadier stance than anything else for this kind of money. There are bucket seats, carpeting, an improved Astro Ventilation system that admits outside air through new rectangular instrument panel vent ports, a wide range of long-lasting Magic-Mirror colors, 25 power team combinations and a new larger tire design. Other refinements you can't see make Camaro so quiet that other sportsters wish they had our secret.

1969

I-6 Convertible

The more powerful of the two available straight sixes for 1969 was the Turbo-Thrift, a 250-cube unit developing 155 horsepower. Peak torque came at just 1600 rpm, so drivers using manual shift didn't have to continually row through the gears to get the most from the motor. Zero-to-60 times could be as brisk as 12.5 seconds—tepid by 21st-century standards but lively enough to be enjoyable in the open air in 1969. Four-wheel disc brakes were available with all Camaros, after having been restricted to the Z28 for 1968. With whatever engine, the standard drums were inadequate for anything other than gentle neighborhood driving, so owners who anticipated taking on the twisties usually went for the uplevel discs. This example is in Olympic Gold.

346 LZZ

Camaro
BY CHEVROLET

346 LZZ

346 LZZ

Camaro SS Sport Coupe with Rally Sport equipment and Super Scoop hood.

**Camaro's new Super Scoop.
Step on the gas and it steps up performance.**

We really put our foot in it this time. Brought out a new Super Scoop hood you can order for Camaro SS and Z/28. It opens on acceleration and socks cool air to the carburetor for more power.

Camaro's got a lot of other scoops, too. It hugs the road with the widest tread of any sportster at its price.

It's the only sportster with computer-selected springs.

Bucket seats, Astro Ventilation, full door-glass styling and wall-to-wall carpets are all standard equipment.

With the SS version, you get all this *plus* a big V8, wide oval tires, power disc brakes and a special 3-speed with floor shift.

Your Chevrolet dealer's got the whole story on how the Hugger scoops the competition. Stop in.

See for yourself.

And step on it.

CHEVROLET

Putting you first, keeps us first.

1969

SS Coupe

For 1969, the Super Sport package cost $295.95 and found nearly 35,000 takers. Chevy had a pair of 350-cid V-8s this year, at 255 and 300 horsepower; the latter was available only with the SS option. Standard SS striping was subdued, beginning low on the front fenders, just behind the headlight surrounds, and tapering up and back as a subtle speed line that ended just ahead of the door handles. The pattern was available as an option with any Camaro save the Z28 and Pace Car Replica. The dual-quad unit hood trim, which had been reserved for 396-cid SS cars in 1968, was standard with all SS Camaros in '69.

Z is for "Zap!"
Translation: a 302 V8 with mechanical lifters, hi-performance cam, aluminum intake manifold, Holley 4-barrel.
Plus: multi-leaf rear springs, heavy-duty shocks, new white-lettered tires on 15 x 7 wheels.
And a Hurst shifter for the 4-speed.
While you're at it, why not add the new 'Vette type 4-wheel disc brakes?
By now you know the mean streak isn't just painted on—it's built in.

CHEVROLET

Putting you first, keeps us first.

We've got a mean streak.

Z/28 Camaro.

1969

Z28

The '69 Z28 didn't have to come in RS guise, of course, as the Z package could be ordered as a stand-alone option. Z28s had mandatory 4-speed Muncie manuals, ensuring that this was a driver's car, rather than just a car owned by a driver. Advertising noted a "mean streak": 60 mph came up from rest in 7.4 seconds, and the Z could run the quarter in 14.8 seconds at 101 mph. Much of that power came from a four-barrel Holley carburetor, an aluminum high-rise intake, heavy-duty radiator, and solid lifters. A 12-bolt rear axle gave efficient handling of the car's 290 pound/feet of torque at 4200 rpm. Need to stop in a hurry? Front discs (or optional disc all around) did a serviceable job. Handling was unusually precise and was aided by quick, communicative power steering. The note of the dual exhausts was tweaked at the factory to give an intimidating baritone growl. Z28 production for 1969 was 20,302, a healthy increase over the 7199 of the year before.

302

Z/28
10 112

Z/28
10 112

Z/28

1969

ZL-1

When Illinois Chevy dealer Fred Gibb heard that Camaro was planning a Cobra fighter for 1969, he asked for 50 such supercars. What he and a few other dealers got (total production was 69) was a Camaro SS stripped of its insignia, and its 396-cid V-8 swapped out for Chevy's mighty ZL-1, an aluminum-block 427 that produced at least 500 horsepower. With factory-added suspension and transmission upgrades, a ZL-1 could zip from zero to 60 in 5.3 seconds and run the quarter-mile in 13.16 at 110 mph. (Race-modified ZL-1s could turn the quarter in the low 10s.) The ZL-1 was a COPO (Central Office Production Order) item rather than the familiar RPO (Regular Production Option). Total price, with $4160 for the ZL-1 package alone, was $7200.

NUMBER
TEAM

USA-1
GOODYEAR

GOODYEAR

GOODYEAR
WIDE TREAD

like Bill Jenkins, . . .

Dick Landy, Shirley Shahan and Dave Strickler. All you have to do is ask the professionals why they choose Cragar Mags. They'll tell you it's the best custom wheel on the road today. Or on the strip. In fact, Bill runs Cragar Mags on his street machine too! His choice for going home is Cragar's new One-Piece "Swinger" wheel. It comes in a light or dark center and rim diameters of 14 and 15 inches. See your speed shop, tire dealer or auto accessories store for Cragar Custom Wheels and Accessories. Or ask Grumpy — he'll put you straight on Cragar Mags.

Grumpy's Toy — '69 Camaro

CRAGAR

JULY 1969 • HOT ROD 9

1969

RS Coupe

The slit-like louvers that covered the headlamps of '69 Rally Sport Camaros slid inboard, via vacuum, when the lights were activated. If the vacuum system failed to work, enough light shone through to allow other cars to see the Camaro—though the Camaro pilot might drive off a cliff. RS models also had headlight washers this year, but the car had to be moving so that pressure and gravity would force the squirts of fluid across the lenses. In addition to unique headlight treatment, RS models mounted parking and backup lights below the bumpers and had thin chrome brightwork around the wheel openings. Many fans of the '69 Camaro have a special fondness for the RS, feeling that it's the right mix of restraint and flash.

RS
MAD 1969
CAMARO

RS
MAD 1969

The Man: Jean-Claude Killy, 3-time Olympic gold medal winner.
The Car: The '69 Camaro SS Convertible with Rally Sport equipment.

All sporty cars are not alike.

vive la différence!

Is Camaro just like any other sportster?

Is Jean-Claude Killy just any skier?

Not a chance. Camaro is as different from other sportsters as Killy is from a snow bunny.

Camaro's got the long, lean look of its sports car cousin, the Corvette. It's got Corvette muscle, too, with a 210-hp 327 V8 standard. Other V8's you can order run all the way up to a 325-hp Turbo-Jet 396.

Camaro isn't called "The Hugger" for nothing. Its wide stance and nimble wheelbase make for cat-quick maneuverability.

Inside Camaro, you think you're in a luxury car with wall-to-wall carpeting and supple vinyl seats.

To make your Camaro even more personal, add a three-speed Turbo Hydra-matic, 4-wheel power disc brakes, an AM-FM radio, you name it.

No, Camaro's no ordinary sportster. It's the personal car that's everything you've been looking for . . . and then some.

Magnifique!

Putting you first, keeps us first.

Catch Jean-Claude on TV. One-hour ABC special runs Monday, January 13, 9 PM (EST). Thirteen-week CBS series begins Sunday, January 5, 5 PM (EST).

1969

RS Convertible

Over the course of the 1969 model year, Camaro offered 14 different engines—not all at once, mind you, but always in combinations that gave buyers plenty of performance choice. New for '69 was a 307-cid V-8, a 283 block fitted with a 327 crank. Horsepower was rated at 200, so at a glance the 307 seems to be an unimpressive performer. The surprise is in the torque rating: an impressive 300 pound-feet at 2400 rpm. The motor was a particularly nice mate for the RS package, which successfully negotiated the lines between performance and economy, smart good looks, and driving ease. Because of the torque, and a light back end, the 307 didn't like snow but handled well in rain if the trans was kept to a low gear.

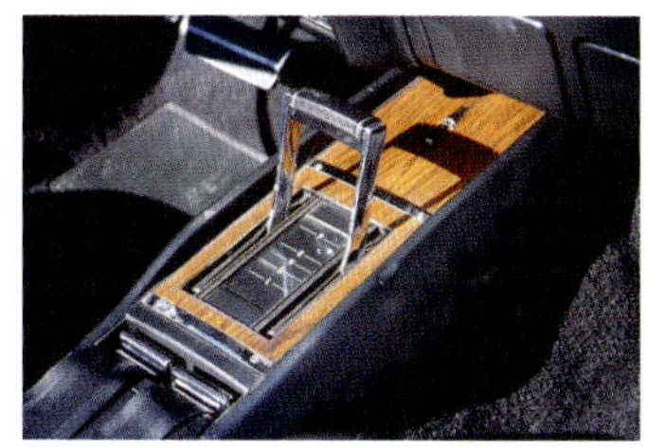

RS
307

Camaro SS Convertible
Camaro Sport Coupe with Rally Sport
Camaro Convertible (V8) 12467 (6) 12367
Camaro Sport Coupe (V8) 12437 (6) 12337

1969

327 Convertible

For the early part of 1969 Camaro production, the standard V-8 was the 327. It was shortly phased out, however, and replaced with a new 307-cid small block. (Because the 327 isn't listed in Chevy literature from later in the model year, the motor continues to be a source of some confusion.) A '69 Camaro with the original 327 isn't as scarce as hen's teeth, but it does excite added collector interest. When mated to an automatic transmission, as on this Monza Red convertible, the 327 produced 210 horsepower. Back seats were narrower in the ragtops than in the '69 hardtops, to allow room for the folding mechanism. "Hockey stick" fender striping was a $15.80 option. Basic red-painted wheels, dog-dish hubcaps, and redline tires ($88.60) completed a handsomely spartan look.

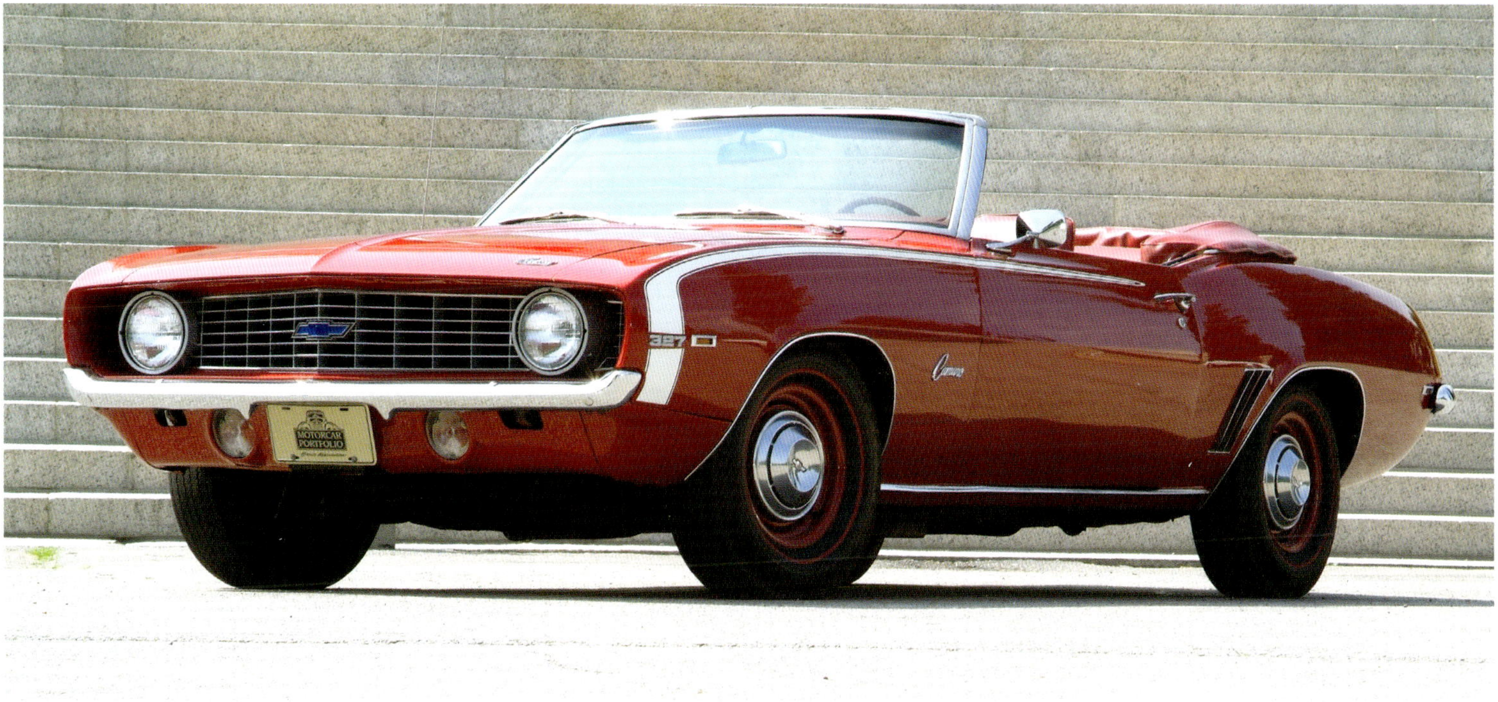
327
MOTORCAR PORTFOLIO
Camaro

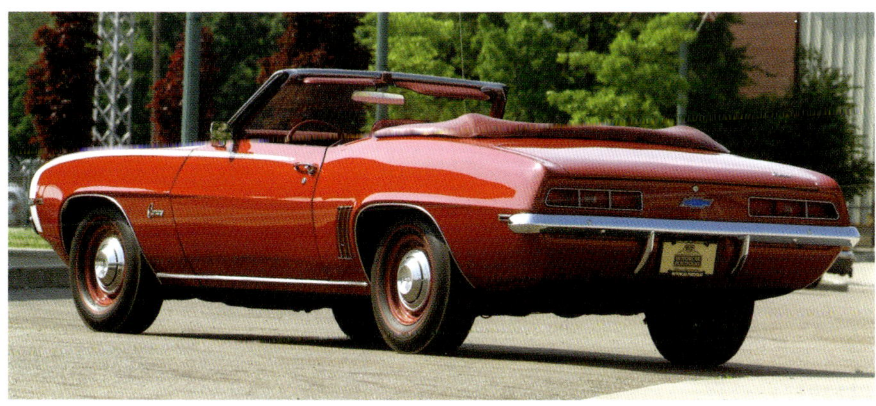

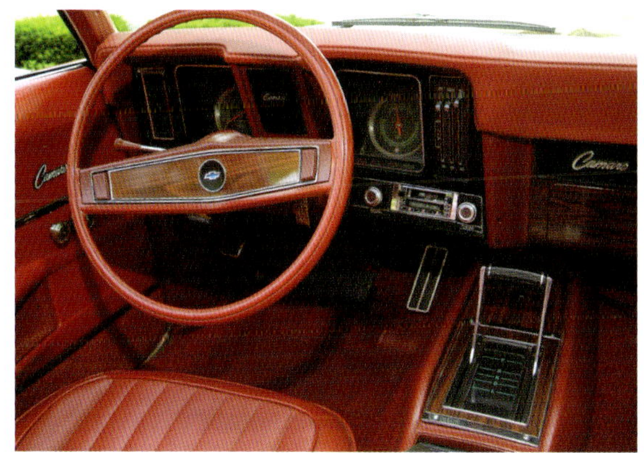

Camaro

The Hugger, Camaro SS Coupe with Rally Sport equipment.

What the younger generation's coming to.

The 1969 Camaro is closing the generation gap. Fast.

Some parents are even asking to borrow their kids' Camaros.

And some kids are actually letting them.

Camaro's secret is its Corvette accent. Standard bucket seats. V8's up to 325 horsepower. And Camaro's the only American car besides Corvette that offers 4-wheel disc brakes.

Camaro's got a lot more going for it, too. Like this SS version that comes with a big V8, power disc brakes, beefed-up suspension, a special floor shift and wide oval tires. And with the Rally Sport package, you've got the only sportster at its price with out-of-sight headlights.

But don't think for a minute that we won't sell you a Camaro if you're over thirty.

After all, it's not how young you are.

It's how old you aren't.

Putting you first, keeps us first.

See Olympic Gold Medalist Jean-Claude Killy, Sundays, CBS-TV. Check your local TV listings.

1969

396 SS Coupe

SS came in two V-8 flavors for '69: 350 and 396. The 350 developed 255 or 300 horsepower, while the 396 dished out 325, 250, or 375. For the record, although muscle-car sales outlets occasionally advertise a "327 SS," no such Camaro was ever produced by Chevrolet for 1969. Your eyes may tell you otherwise, of course, because it's not impossible for the naïve or unscrupulous to do convincing counterfeits of badging, and engine and chassis numbers. Thus, every 327 SS model still running around started life as another model, often a base Camaro. The "hockey stick" stripe scheme was standard on SS. This 396 SS is finished in Hugger Orange and has a black vinyl roof, plus Positraction and a 4-speed Muncie transmission with a Hurst shifter.

SS
396
Radial T/A
BFGoodrich

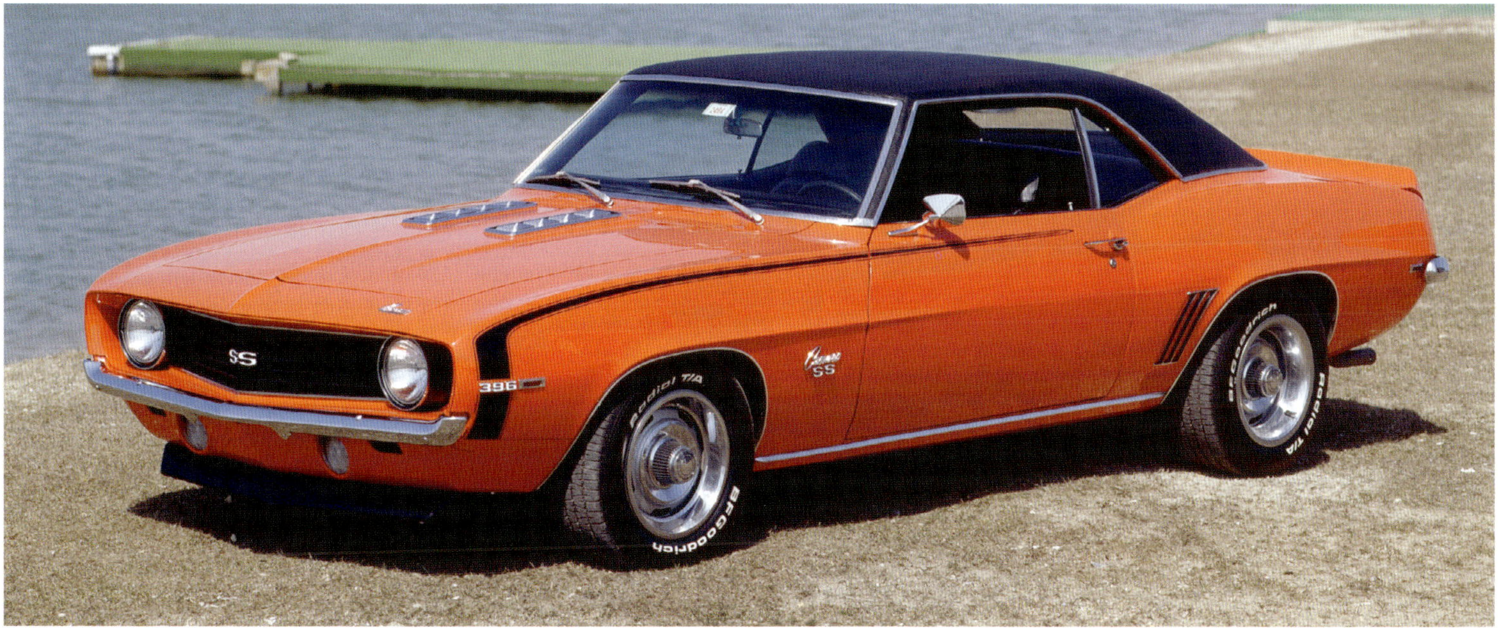

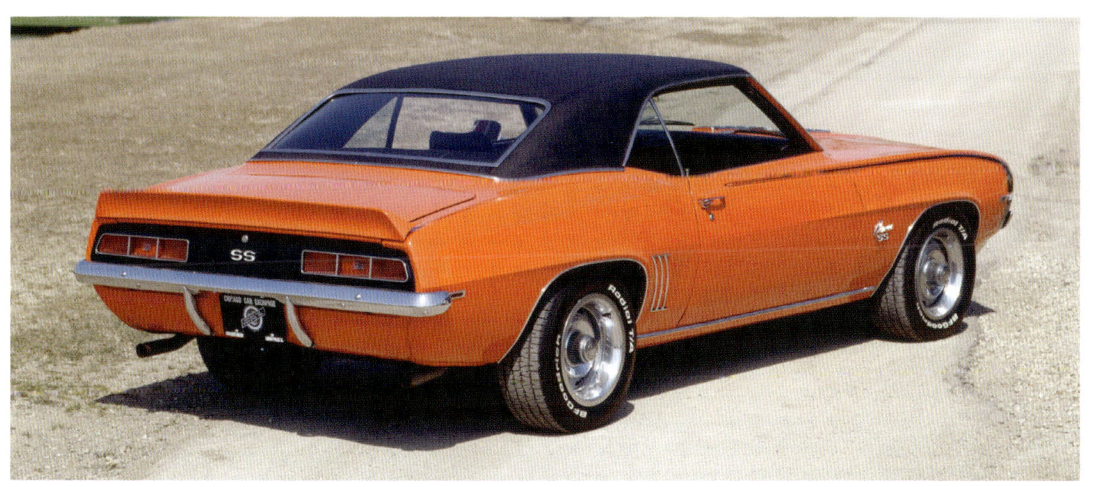

SS
Radial T/A

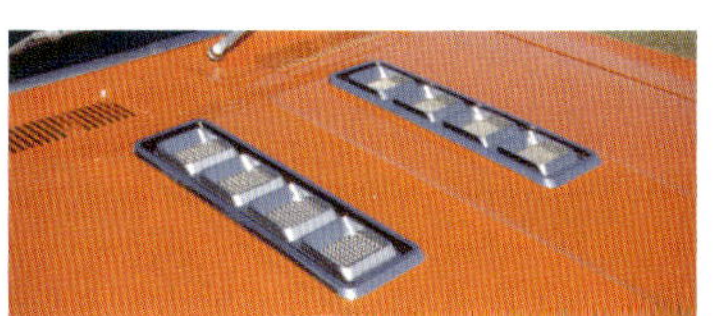

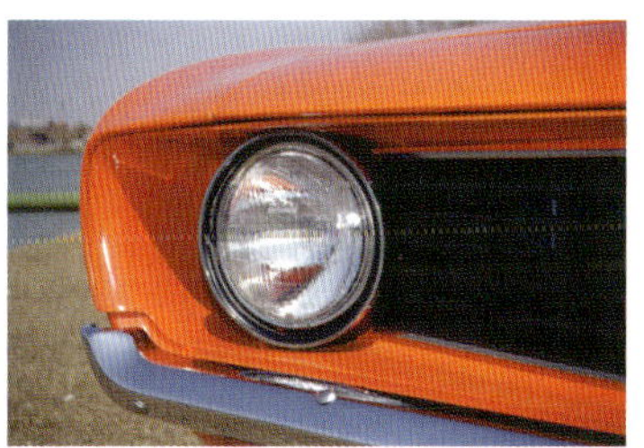

Mary Anne Zovko, University of Pittsburgh, Eastern College Athletic Conference
Katrina Hampel, University of Miami, Major Independent College
Nancy Chase, Iowa State University (Ames), Big Eight
Judith Ann Kostler, Clemson University, Atlantic Coast Conference
Brenda Shrom, University of Delaware, Eastern College Division
Suzi Yung, Ohio State University, Big Ten
Cindy Demarest, Western Michigan University, Mid-American Conference
Nancy Garretson, University of Cincinnati, Missouri Valley Conference
Kathy Loghry, Oregon State University, Pacific Eight
Karen Cope, University of Mississippi, Southeastern Conference
Grace Roselli, College of William and Mary, Southern Conference
Trinette Hudson, University of Nevada, Western College Division
Ellen Cullaton, University of New Mexico, Western Athletic Conference
Barbara Specht, Texas Tech, Southwest Conference

Vote for the queen of your choice.

1969 Camaro SS Convertible with Rally Sport equipment. Putting you first, keeps us first.

The National Collegiate Athletic Association and Chevrolet invite you to select one of these lovely coeds to reign as Queen during the 1969 celebration of college football's 100th anniversary. Each girl is already a queen. Each was elected in sectional contests to represent the NCAA's eleven major conferences, major independents and College Division.

Now, it's up to *you* to pick one to be the National Centennial Football Queen.

The Queen and her two Princesses will be announced and crowned during half-time ceremonies at the Shrine East-West Game on Saturday, December 28. Be sure and watch this exciting moment live on ABC-TV.

The Queen will receive a $1000 scholarship from the NCAA.

She'll also be given this fit-for-a-queen 1969 Camaro Super Sport Convertible to enjoy during her reign.

Both Princesses will receive a $500 NCAA scholarship.

Don't miss your chance to choose a Queen. Mark your ballot and vote *now*.

CONTEST RULES

1. On the official ballot, print the name of the girl you think should be the College Football Centennial Queen.
2. Additional votes may be entered on a plain sheet of paper. Only one ballot per envelope or post card, please.
3. Ballots must be received by Dec. 18

CHEVROLET

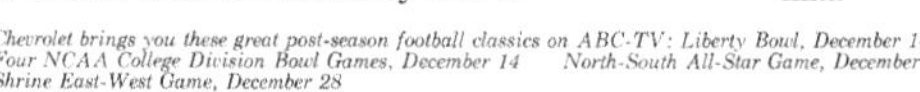

Chevrolet brings you these great post-season football classics on ABC-TV: Liberty Bowl, December 14 Four NCAA College Division Bowl Games, December 14 North-South All-Star Game, December 25 Shrine East-West Game, December 28

OFFICIAL BALLOT

My choice for the College Football Centennial Queen is:

____

Mail to:
Centennial Queen Contest
P.O. Box 1010
Detroit, Michigan 48232

1969

396 SS Convertible

For 1969, only Camaros with the SS package could be ordered with the top 396 V-8, the 375-hp L78. Because 59.3 percent of the weight of a 396 Camaro was above the front axle, the lightly loaded rear axle struggled to gain the traction needed for quick, sure getaways. Contemporaneous road testers complained of wheel hop and back-end judder—a clear squandering of some of those 375 horses and 415 pound feet of torque. Stoplight jockeys and legit racers who used their noodles fitted their 396 Camaros with aftermarket traction kits, which kept the rear wheels planted. Very few of the year's 396s were convertibles. This example has the new-for-'69 $95 gauge pack with a round tachometer in the main instrument pod, plus auxiliary gauges mounted at the console. This car also has a power top, a $52.70 option.

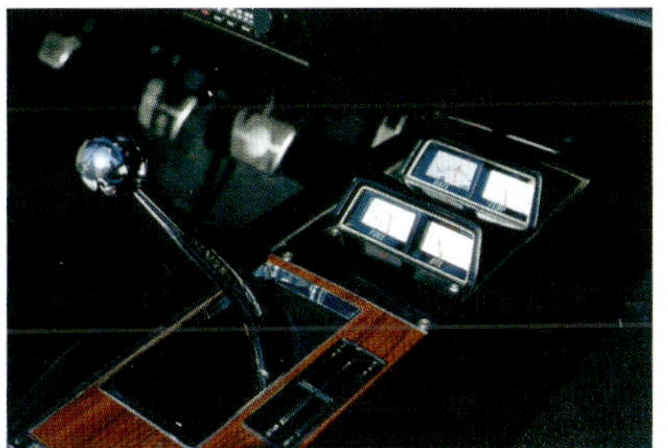

Camaro SS Convertible with Rally Sport equipment and new air intake hood.

Why is the Camaro the pace car again? Because it's *the Hugger.*

Camaro has been named the Official Pace Car in the Indianapolis 500 for the second time in three years—a 50-year record! If you haven't driven the Hugger, take a hint from the guys at Indy. Maybe they know something you don't.

Camaro SS has what it takes. Again this year, it'll lead the pack at Indy.

Engine choices start with a 300-hp 350-cu.-in. V8 and run up through a 325-hp 396-cu.-in. job. There's even a new hood you can order with a super scoop intake that opens on acceleration, ramming cooler air into the engine for more power.

This Hugger offers the widest tread of any sportster at its price. It comes on strong with wide oval tires on 14 x 7-inch-wide wheels and a beefed-up suspension.

The transmission is a special 3-speed floor shift. For those who want still more, there's a 4-speed available with a Hurst shifter.

Indy's tough. So's Camaro SS.

When it comes to pacesetting, Camaro knows its way around.

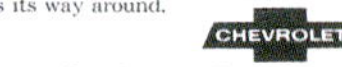

Putting you first, keeps us first.

1969

Indy Pace Car Convertible Replica

Camaro paced the 1969 Indianapolis 500 and celebrated by issuing 3675 replicas; RPO for this option was Z11. The cars ran with the top 350-cid V-8, rated at 300 horsepower. A 3-speed manual was standard, although the example seen here runs with an optional automatic. Bodies were Dover White with Hugger Orange stripes; large, wordy door decals were shipped in the trunk and were applied at the owner's discretion. The cloth upholstery was done up in orange houndstooth (a houndstooth pattern—in black, orange, or the rare yellow—was the only cloth option for Camaro this year). Rally wheels and cowl induction hoods were part of the replica package. Other options that jibed with the official Indy Pace Car included a console, front shoulder belts, rear spoiler, AM radio, power steering, and Positraction.

The 1969 Camaro SS keeps tough company.

For two out of the last three years it's been chosen the official Indy 500 pace car. Check one out and you'll know why.

Engine choices start with a 300-hp 350 V8 and work their way up. For more power, a special new intake hood is available. We call it Super Scoop. It opens on acceleration, ramming huge gulps of air to the engine.

Additional credentials: beefed-up suspension, white-lettered wide ovals on 14" x 7" wheels, power disc brakes and a special transmission with floor shift.

It takes a lot to get this crowd started. That's why Camaro SS sets the pace.

For competition to follow.

CHEVROLET

Putting you first, keeps us first.

Camaro SS...and friends.

Camaro SS Convertible with Rally Sport equipment and new Super Scoop.

1969

SS Z10 Coupe

The '69 Z10 "Indy Pace Car replica" coupe has been sighted by few and denied by many. Is it real? The answer is yes—with caveats. No Camaro hardtop appeared at the '69 Indy race in pace-car trim, and the little-known Z10 option never was linked to Indy in Chevy literature. Confusion exists, in part, because the Z10 shared its RS/SS origin with the official ragtop Camaro pace car and picked up the Dover White/ Hugger Orange paint scheme, as well. Chevy's intent was to sell the Z10 in the South and Southwest only. The cars were assembled at Chevy's Norwood, Ohio, plant; 200 to 500 examples were produced (accounts vary). The typical Z10 ran with the 300-hp 350 V-8; the 396 was optional, and of those, most were the milder, 325-horse iteration, though some Z10s got the 375-hp version.

Camaro SS
BFGoodrich
Radial T/A

467 NED

48 1969 CAMARO V8 290 HP V8 TURBO FIRE 302 302 CU. IN. ENGINE W/4 BBL. CARB.

ENGINE 302 Cu. In.

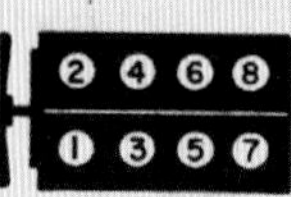

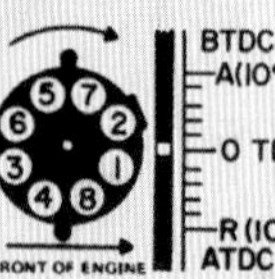

Firing Order 1-8-4-3-6-5-7-2
Mech. Lifters—Engine at normal temp. and running
In. .030″
Ex. .030″
Comp. Ratio 11.0-1
Comp. PSI (min) 190
(max. var.) 20

BATTERY 12V Neg. Grd.

Standard	Y55	45AH
Optional	R69	70AH

Cranking Voltage (min) 9.0v

CHARGING SYSTEM

	Standard	W/Air Cond.
Output (min)	30A	30A
Operating Voltage	13.5-15.2v	13.5-15.2v
Regulator Cover Off (measured at battery)	14.2-14.6v	14.2-14.6v

IDLE
Engine Vac. 16″-20″
RPM—automatic load leveler motor—inactive
W/AIR CLEANER REMOVED—Plug eng. vac. supply port.
Manual Trans. 900 W/Air Cond. off

DWELL (Degrees)
At Idle 30 (28-32)
Variation 2 (between idle and 1500 RPM)

IGNITION TIMING (Degrees)
Dist. Vac. Line Disconnected and Plugged
Manual Trans. 4 BTDC @ Idle RPM

IGNITION ADVANCE (Degrees @ 2500 RPM)

Cent. & Vac.	Cent. Only
36½-40½	22½-24½

IDLE MIXTURE
Air Cleaner Installed—Vac. Hose Connected
With idle speed RPM set to specifications, and mixture screws backed out 3 turns from lightly seated position. Turn mixture screws equally as required to obtain max. idle speed RPM. Adjust carb. idle speed screw to obtain specified idle speed RPM setting, turn each mixture screw clockwise (lean) to obtain a 20 RPM engine speed drop and then counterclockwise (rich) ¼ turn. Readjust carb. idle speed screw as required to obtain specified idle speed RPM setting.

FUEL PUMP
Press. 5-6½ PSI @ 450-1000 RPM
Vol. 1 pt. 30-45 sec. @ Idle RPM
Filter (carb. inlet nut) Replace element every 12,000 miles or 12 months.
Filter (fuel tank) Strainer—Service as required.

SPARK PLUGS

	Standard	Colder
	ACR43	ACR42-1C
Gap	.035″	.035″
Torque	25 Ft/Lbs.	25 Ft/Lbs.

CARBURETOR
Manual Trans. Holley 4150-3923289-(R4053A)
THERMO AIR CLEANER
Engine off and underhood temp. below 85°F, Snorkel Passage should be open—Heat off position.
Engine at idle speed and underhood temp. below 85°F, Snorkel Passage should be closed—Heat on position.
With underhood temp. 85°F-128°F, Snorkel Passage control damper door should begin to close.
Underhood temp. above 128°F, Snorkel Passage should be open—Heat off position.
With a min. of 9″ vac. applied to diaghram assy., damper door should completely close Snorkel Passage.

DISTRIBUTOR

	Manual Trans.
Delco	1111480
Rotation	C
Spring Tension (oz)	19-23
Gap (inch) New	.019
Used	.016
Dwell (Degrees)	30 (28-32)
Variation (Degrees)	2
Condenser Capacity	.18-.23 MFD

MECHANICAL ADVANCE
1111480

Dist. RPM	Dist. Deg.
625	0
1100	11
2200	16

VACUUM ADVANCE

In. Vac.	Dist. Deg.
8	0
15½	7½

IGNITION COIL (ohms @ 80°F)
Delco 1115293
Pri. Res. 1.77-2.05
Sec. Res. 3000-20,000
Test Set Line 8
Ignition Resistor
Loom 1.35

IGNITION CURRENT
Engine Stopped 4.0A
Idling 1.8A

SECONDARY RESISTANCE 3.0 min.

STARTING CRANKING CIRCUIT RESISTANCE
Insulated Circuit
Bat. Pos. Post to Starter Motor Terminal of Solenoid—.4v
(Pos. bat. cable only— .2v)
(Solenoid contacts— .2v)
Ground Circuit
Bat. Neg. Post to Starter Motor Housing—.2v

STARTER FREE RUNNING CURRENT DRAW
Delco 1108367
53-69A (includes solenoid) @ 9.0v RPM 6400-8600

SOLENOID CURRENT DRAW
Hold in Windings 14½-16½A @ 10v
Both Windings 41-47A @ 10v

SOLENOID PULL IN VOLTAGE
SW (S) Term. of Solenoid and Ground—7.7v (min)

SOLENOID CONTROL CIRCUIT
Bat. Term. of Solenoid to SW (S) Term. of Solenoid—3.5v (max)

FORM NO. 690-690-048 **SUN ELECTRIC CORPORATION**—CHICAGO, ILLINOIS 60631

1969

RS/Z28

Chevy's 302 Turbo-Fire V-8 remained exclusive to the Z28 sport coupe for 1969 and was conservatively rated at 290 horsepower. Four-wheel disc brakes had four-piston calipers, as on the year's Corvette. The deck and spoiler rally stripes seen on this RS/Z28 are factory-installed and correct, as they end at the base of the spoiler. Stripes applied later by dealers or owners often cover the narrow portion of the trunklid that's visible beneath the spoiler. Z28 options included a cowl-induction hood, fold-down rear seat, tilt wheel, and an 8-track tape player.

MEET THE
MEAN ONES!
YENKO/SC
USA 1
sYc
Chevelle
YENKO/SC
YENKO/SC
sYc
sYc
THE YENKO 427 S/C SUPER CARS
You've waited long enough to tame your kind of town and your kind of road--on your kind of terms. Here are the Yenko 427 Super Cars! And they're all GUTS!
SPECIFICATIONS:
. . . 427 CID 450-hp engine with hi-lift cam and solid valve lifters.
. . . Hi-rise manifold with 800 CFM Holley four-barrel carburetor
. . . Four-speed close-ratio Muncie transmission with special Hurst shift linkage (dual gate floor-mounted linkage with Turbo Hydramatics)*
. . . Power disc front brakes/15x7 wheels (14x6 on Novas) WT GT tires
. . . 4.10:1 positraction rear axle with special heavy duty ring, pinion and axle shafts
. . . Reversed air-breathing hood scoop with built-in sealed plenum chamber and throttle operated solenoid valve*
. . . Computer tuned heavy duty suspension including special shocks, springs and extra H.D. stabilizer bar
. . . Front under-grille spoiler
. . . 450 hp @ 5000 with standard exhaust configuration
. . . One piece Aluminum cast Mag Wheels optional at extra cost with Yenko emblem
STANDARD APPEARANCE FEATURES:
. . . Custom YENKO side and hood striping . . . Unique headrest styling . . . Distinctive 427 & YENKO emblems . . . Transistorized tachometer . . . 140 mph speedometer* . . . Rear deck spoiler*

1969

Yenko SC Coupe

Super Camaros from Pennsylvania Chevy dealer Don Yenko were hand-built for 1967 and '68, with outside contractors swapping out engines and performing other performance mods. But by 1969, Chevy caught on to the PR value of Yenko's activities and gave full factory support. Because Yenko was to be the manufacturer of record, Chevy Division could slide around the 400-cid limit imposed on Camaro by GM. So, the '69 Yenko Camaro (COPO 9561) ran with Chevy's 427 V-8, rated at 425 hp—uprated by Yenko to 450, to match NHRA test results. Mandatory trans was Muncie's M21 4-speed manual. The car cost about $4400, considerably less than the $7200 demanded for the special-order ZL-1 427. Total Yenko Camaro production for '69 was 198.

1970

Z28 Coupe

The aggressively handsome 1970 Camaro retained the 108-inch wheelbase but didn't share a single body panel with its first-generation predecessor. The unified, long-hood/short-deck body, fronted by an aggressively outthrust nose, excited buyers. The standard Z28 engine no longer was the fine 302, but Chevy's LT-1 350, rated for the Z at a thumping 360 horsepower. Because the new Camaro wasn't ready for its public introduction until February 1970, production was dramatically truncated, with just 124,889 Camaros produced for the model year. Of that number, a mere 8733 were optioned as Z28s. Hugger Orange, with white rally stripes, was a popular exterior color combination.

Z28

1970 VIRGINIA
404·999
Radial T/A
BFGoodrich

1970 VIRGINIA
404·999
Radial T/A
BFGoodrich

New Camaro Z28

Separates the men from the toys.

Remember when you were a kid and you put a lot of trick stuff on your bike to make it look like something it wasn't?

A lot of so-called "sporty cars" still operate that way.

But not this one.

The new Camaro Z28 is as good looking underneath as it is on top.

With a 360-horse Turbo-Fire 350 V8. And with a Hurst shifter that comes along for the ride when you order the 4-speed.

Then there's the suspension that lets you feel the road without feeling the bumps. And the quick ratio steering. And the special wheels with the F60 x 15 tires. And on, and on, and on.

But don't just take our word for it. Pick one up at your Chevy dealer's Sports Department and take it for a road test.

You'll see we're not kidding around.

Putting you first, keeps us first.

1970

Z28 Coupe

At $573, Camaro's 1970 Z28 package wasn't inexpensive, but by any standard it was a high-performance bargain. The Z's heart and soul was the 350-cid, 360-hp LT1 V-8, an engine that was as sophisticated as it was potent: solid lifters, a hot cam, aluminum pistons, oversized valves, and a 780-cfm Holley four-barrel carburetor. Compression was 11.0:1. So-called "mandatory extras" were a Muncie 4-speed manual transmission with Hurst shifter ($206) and Positraction ($44) with 3.73:1 gearing (a 4.10:1 gear set was available). The new motor produced enough low-end torque for Camaro to offer its first-ever automatic transmission option, the Turbo 400. The Z28 was now indisputably a serious road car that could stand with many of the best from around the world.

Z28

Z28

1970

Z28 Hurst "Sunshine Special" Sunroof Coupe

Camaro eliminated its convertible this year—a shock to open-air fans but an opportunity for a novel experiment by aftermarket manufacturer Hurst: three examples of the Sunshine Special Z28. Because the front bumper isn't full-width, and round parking lights flank the headlamps, the car's RS accoutrements are obvious. (Standard Z28s had a full-width front bumper.) The Special's canvas roof insert cranked into the headliner, accordion-like, to allow the sun in. The mechanism was never mass-produced for Camaro, but it was adopted for Chevy's Nova in 1972 and called the Skyroof. Although changes in SCCA rules disqualified Hurst's grinning front spoiler for production use, the three-piece rear spoiler design was adopted as an option for the year's production Camaros. A T-handle Hurst shifter in a slotted gate helped drivers make the most of the 350-cid, 360-hp V-8.

HURST
Z28

864 XHS
BFGoodrich
Belted T/A

POLYGLAS GT F60-15
GOODYEAR
New Polyglas® GT
Goodyear's widest street tire yet
Eight inches of tread width puts more rubber on the road. 60-Series — only 60% as high as it is wide. With big, white Goodyear letters on the sidewall. Standard or optional on several 1970 supercars. Only one problem. It's so wide it may not fit your car. Check your car's specs before you buy. If it doesn't fit — maybe you need a new car.
GOODYEAR

1970

SS/RS Coupe

'70 Camaros ordered with the RS package came with Euro-style parking lamps that sat level with the headlights above an abbreviated, bladelike bumperette; no bumper crosses the RS grille. Buyers who combined the RS with the SS option drove off with a 350-cid V-8 rated at 300 horsepower. The three-piece spoiler devised by Hurst was an option (it would become standard for 1971). The model year brought a few firsts for Camaro: rearview mirror glued to the windshield, side-impact door beams, in-glass antenna, and concealed windshield wipers (optional). Instruments, even optional ones, now were grouped together in the main pod; none were at the console. The SS/RS seen here is in Shadow Gray.

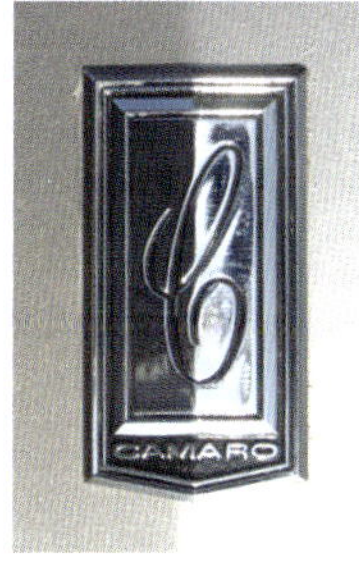

An honest look at driving from the Chevy side of the tracks.

A sports car for the four of you.

Say you've got a wife and two kids. Or you have two Great Danes. Or maybe you have two too many friends. But you want to drive a sports car.

You could buy a Corvette. But good as it is, space is limited to two people.

So, we also have Camaro. With four seats. And we think we're perfectly justified in calling it a sports car.

It certainly qualifies on looks. Our stylists have seen to that. And it performs and handles like a sports car. Our engineers have seen to that.

But nothing we can say can take the place of a test drive.

So why don't you see your Chevrolet dealer. He's got a sports car waiting. For the four of you.

You've changed. We've changed.

1971

Z28 RS

The '71 Camaro differed from its 1970 predecessor in such details as new seats (adapted from Vega) and side marker lights that flashed in sync with the blinkers. Wheel covers got a restyle, and 12 new body colors were added. In a weight-saving move, the windshield glass was thinner than what had been installed the year before. A reduction in hi-po engine compression—due partly to the ramifications of new unleaded, low-octane gasoline—prompted Camaro to list its horsepower ratings in net terms, which allow for power losses in the driveline and from accessories. RPO Z28 brought the 330-hp Turbo-Fire 350, 3.73:1 Positraction, power brakes, heavy-duty radiator, blackout grille, uprated springs, and more. The Cranberry Red example seen here has Muncie's close-ratio "Rock Crusher" M22 4-speed, an option available only with the Z28.

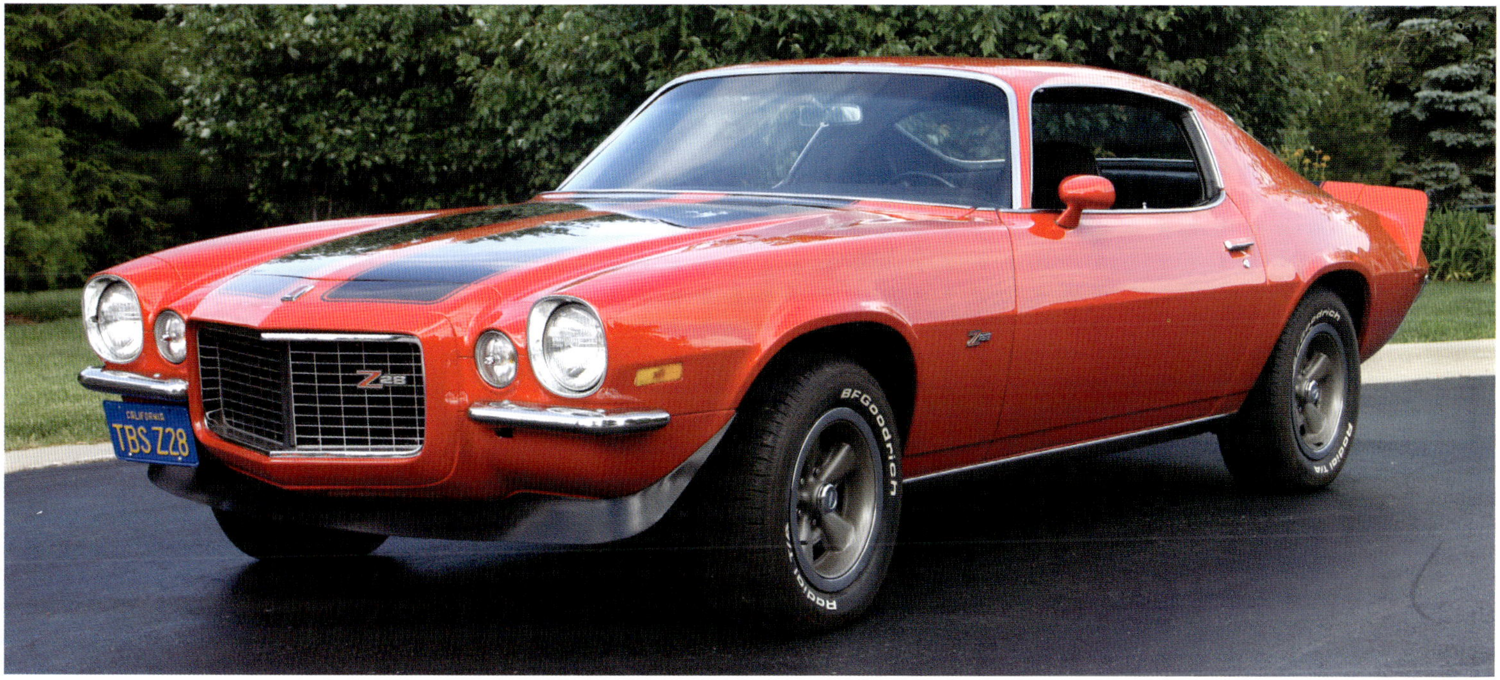
CALIFORNIA
TBS Z28
Z28
BFGoodrich

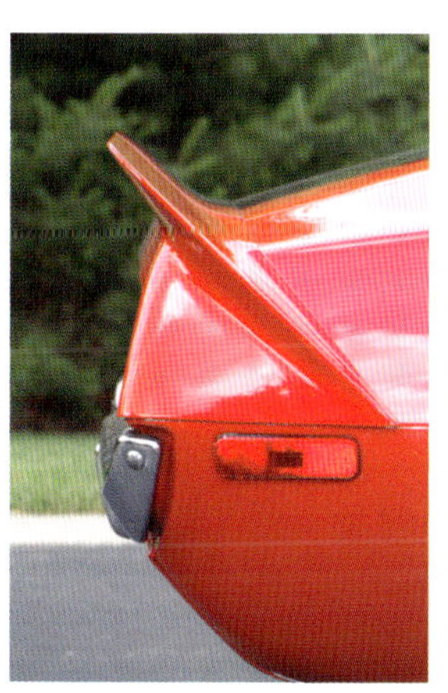

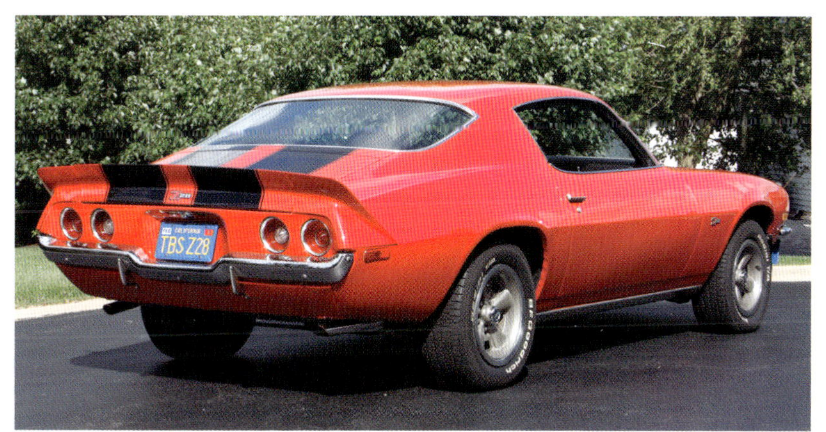

TBS Z28

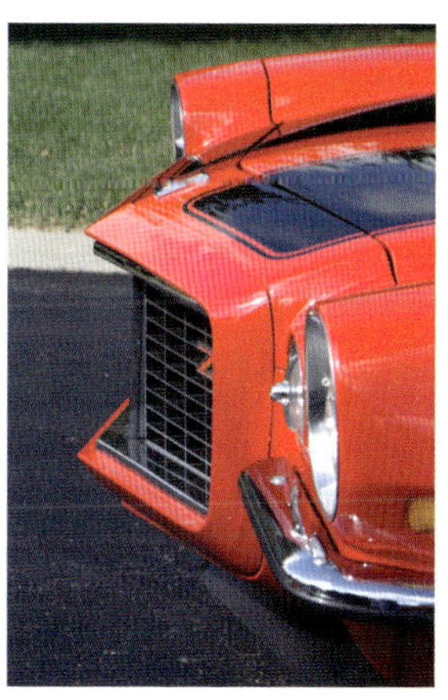

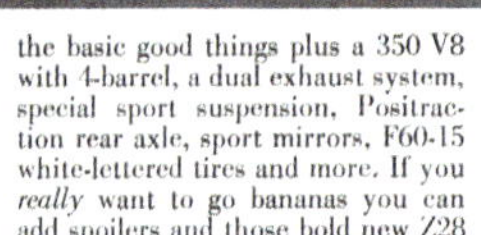

1974

Z28 Coupe

Buyer enthusiasm was sparked this year by the first significant exterior restyle of the second-generation Camaro. Production for 1973 had been 96,751 but leapt to 151,008 for '74. The RS model had gone away, so the Z28 package was mated to the Type LT that had debuted for 1973. For '74, Z28s with the 350-cid, 245-hp Turbo-Fire V-8 could have a Muncie M21 4-speed. Alternatively, Z28 buyers willing to part with $297 could have a heavy-duty Turbo Hydra-Matic. Federal bumper regulations continued to be stringent, but Camaro's new aluminum units met requirements without looking awful. Redesigned taillights were elongated deltas. Model-year 1974 was the last year for the Z28 until 1977. The example seen here is finished in Antique White.

Z28
Camaro

Camaro

Z28

1977

Z28 Coupe

Although Camaro fielded no Z28 when the '77 Chevy line was unveiled, aficionados knew a revival was coming. It was introduced in February 1977, and this time it was no longer an RPO but a full-fledged model. (This Z28 is sometimes referred to as a "1977½," but that designation is strictly informal and doesn't reflect Camaro production reality.) Federal emissions and mileage requirements continued to be a performance bugaboo, so the Z28's 350 V-8 had a net horsepower rating of 165—mild, although not as listless as it may seem. Still, Chevy promotional material for '77 downplayed performance to emphasize the Z28's "outstanding road car" status. A Z with the Borg-Warner Super T-10 4-speed could get the most from the motor. Bronze Metallic with gold accents was a popular combination.

Z28

camaro
Z28
Firestone

Z28
camaro
Z28
Firestone
AUE862

Z28
IOWA 75
AUE862

camaro

1978

Z28 Coupe

Molded urethane body pieces help identify the subtly redesigned '78 Camaro. The fascia was a single piece that incorporated the headlight housings, grille surround, and signal lamps. Bumpers, front and rear, were molded, as well, and the tail lamps were freshened. Z28s got a unique hood, with scoop, plus blackout grille; sport suspension; 4-speed manual transmission; accent stripes; and a variety of decals. The Z28 ran with a 185-horse variant of the hydraulic-cam 350 V-8 (for California, horsepower was 175). *Car and Driver* testers ran the 185-hp Z28 from zero to 60 mph in 7.3 seconds. And a point of trivia: For the first time in Camaro's history, the name for white body paint was just plain White—no more modifiers like Dover and Antique—and "White" it would remain until 1991, when Arctic White arrived.

350

camaro
Z28

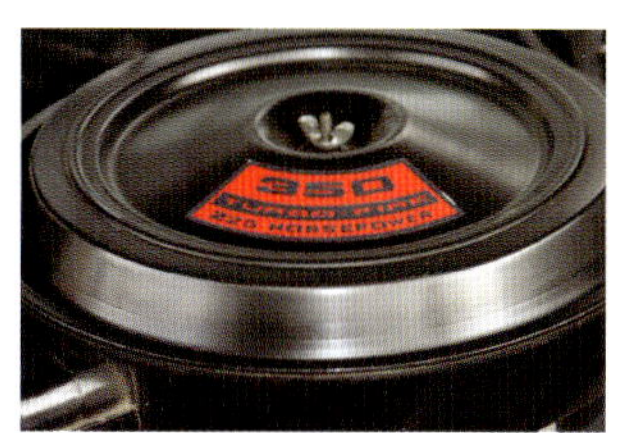

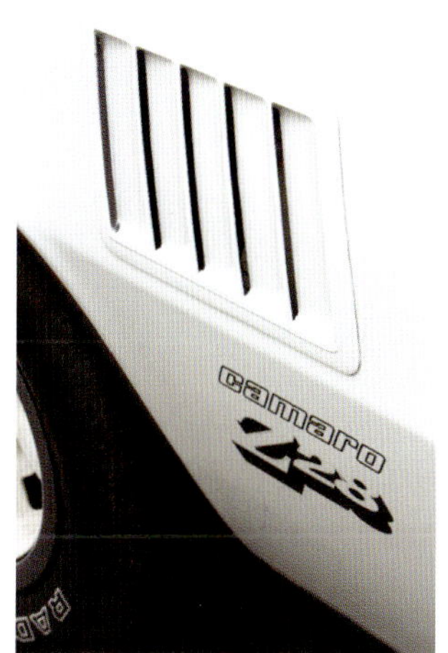

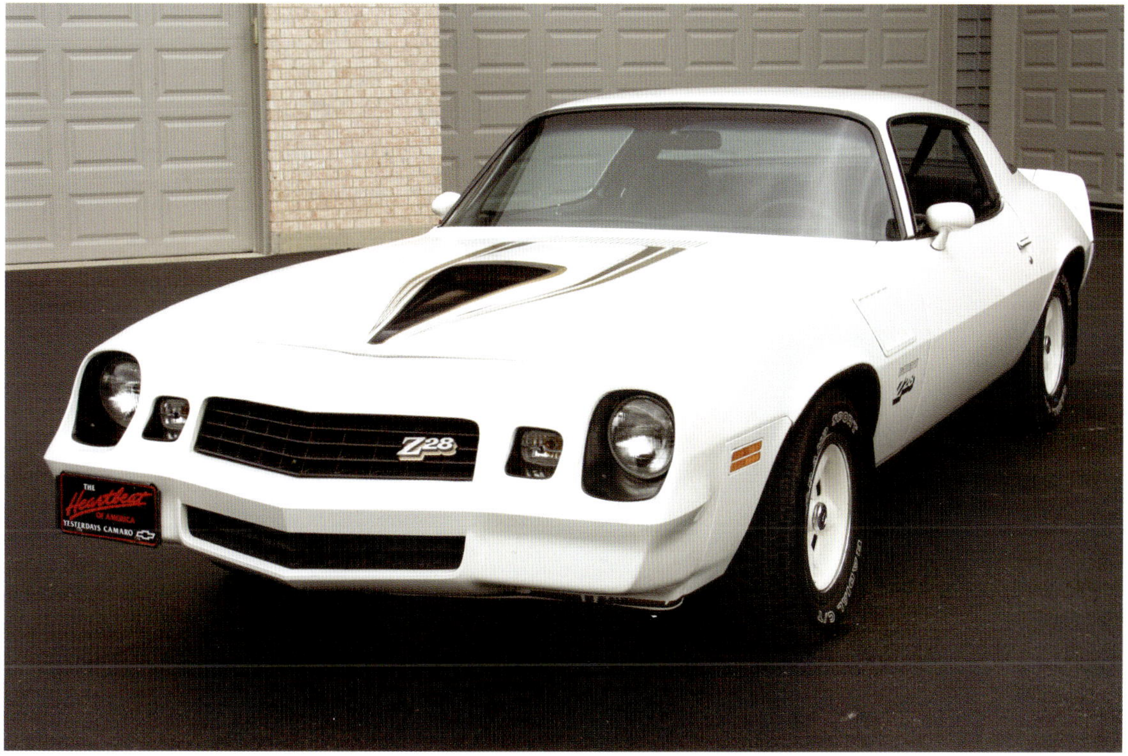
Z28
THE Heartbeat OF AMERICA
YESTERDAYS CAMARO

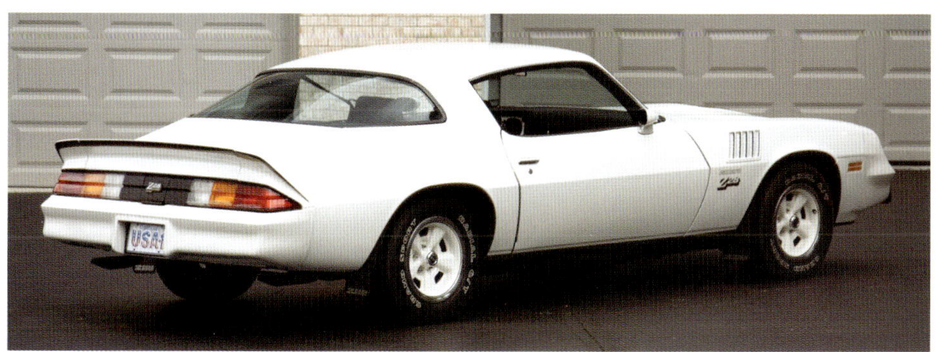

USA1

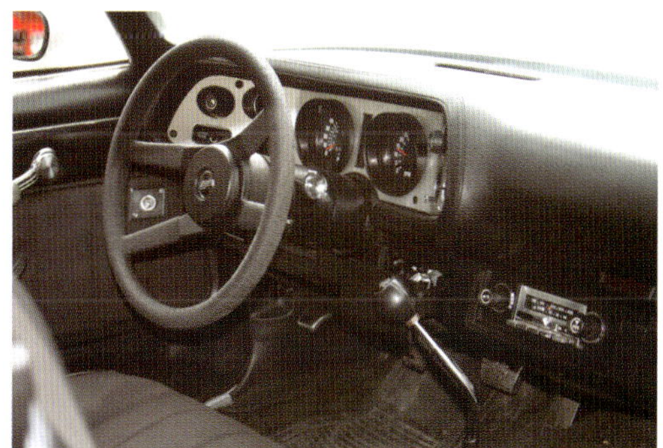

You've seen some of the best drivers in the world driving race-prepped Chevrolet Camaro Z28s in the International Race of Champions.

And maybe you imagined yourself behind the wheel.

For when it comes to hugging a road, a Z28 is truly something special.

We start with a special Z28 Sport suspension system. Then we add things like a 4-speed close-ratio manual transmission connected to an 11-inch high-capacity clutch. Front/rear stabilizer bars, special shocks, power steering, power brakes. To top it all off, we pack a 5.7 Litre (350 Cu. In.) 4-barrel V8 under the hood. (Camaro is equipped with GM-built engines produced by various divisions. See your dealer for details.)

Because a car of Z28's caliber should look as good as it moves, we've given special attention to its appearance. It comes with a rear deck spoiler, new front air dam, bold accent striping, simulated air louvers in the front fenders, sport mirrors, distinctive black finished grille, and more.

Inside, you'll find a special large-rim sport steering wheel, full foam bucket seats, tachometer, voltmeter, temperature gage, an electric clock, and more.

Now stop imagining yourself behind the wheel. Grab hold and take it from Chevrolet—people who know what performance is all about.

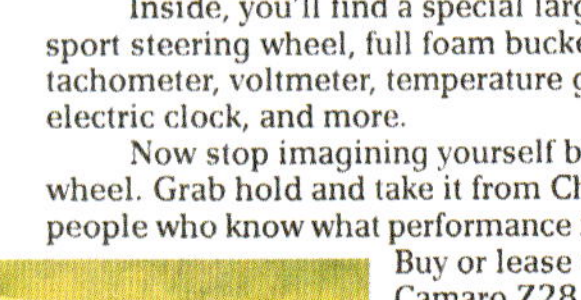

Buy or lease a '79 Camaro Z28. Then go road hunting.

Aluminum wheels shown are available at extra cost. Available white-lettered tires are supplied by various manufacturers.

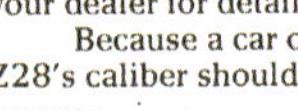

CAMARO. THE HUGGER.

1979

Z28 Coupe

Camaro styling for '79 was essentially carried over from the previous year, though the Z28 picked up a dramatic front air dam that flowed into the leading edges of the front wheel wells. Fender air louvers were another front-end addition. Exclusive bezels adorned the Z's front and rear lights, and the blackout grille continued. However, the "Z28" grille emblem was dropped and replaced by decals. Radio with cassette was a new option; likewise a CB. Output of the 350-cid V-8 was ostensibly downrated this year, from 185 horsepower to 175. The 4-speed close-ratio transmission was an apt mate for the motor—but in California it was an automatic transmission (and just 170 hp) or nothing. In 175-horse trim, the engine developed 270 pound/feet of torque at 2400 rpm.

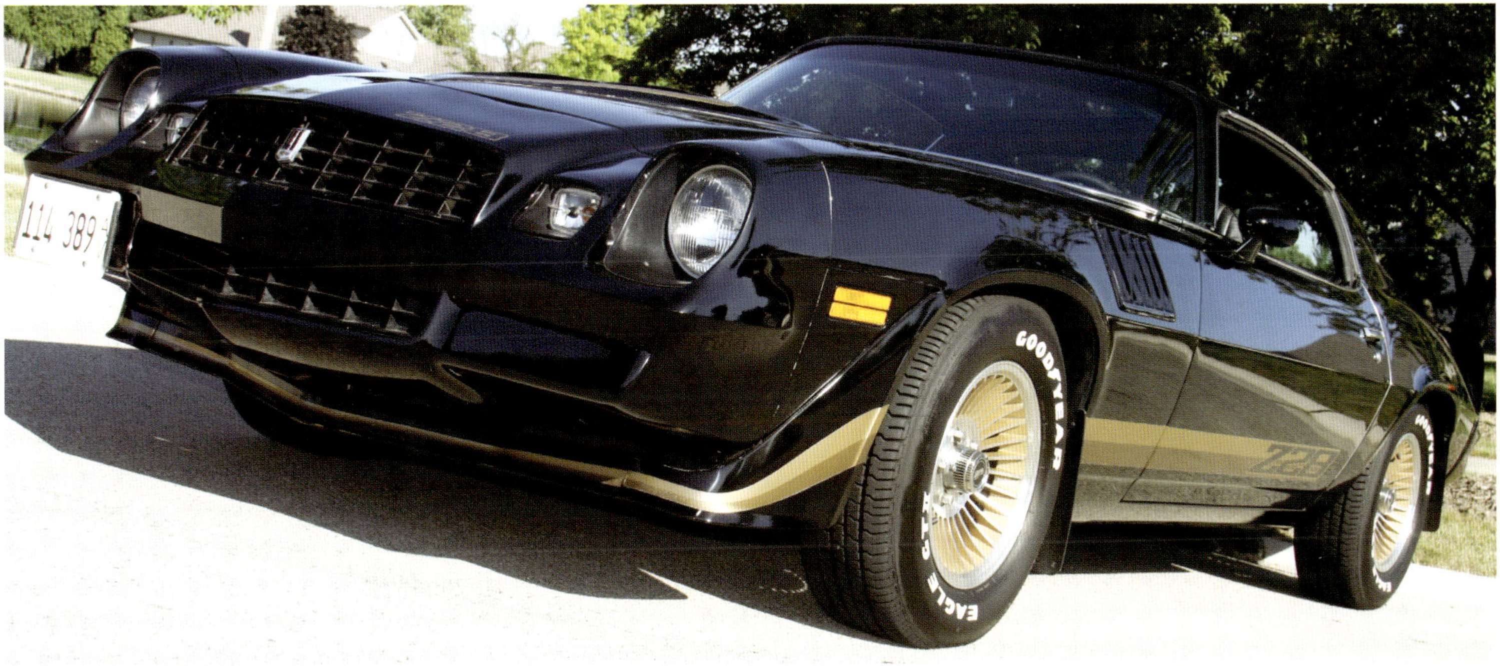
114 389
GOODYEAR
EAGLE GT
Z28

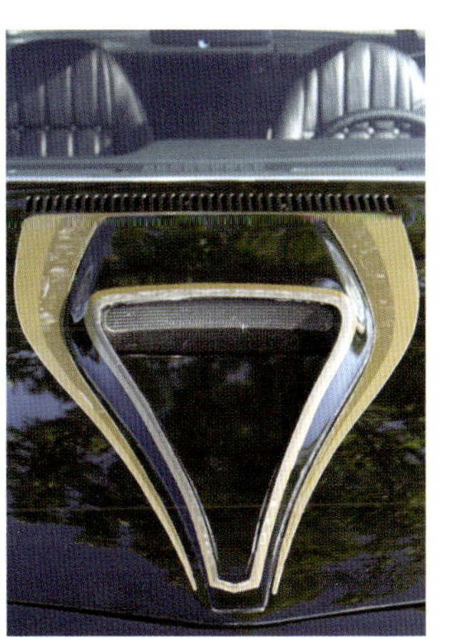

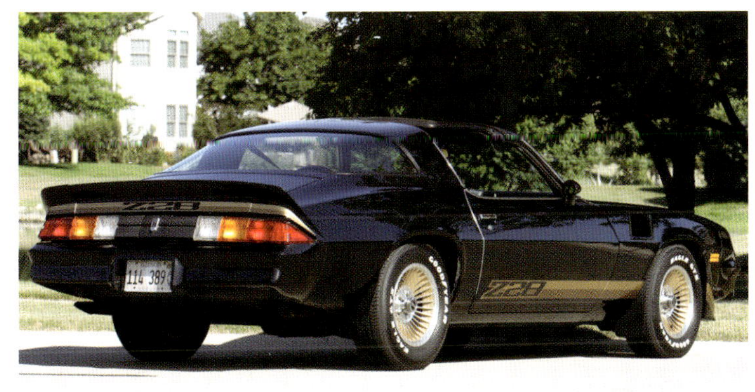

Z28
114 389
Z28

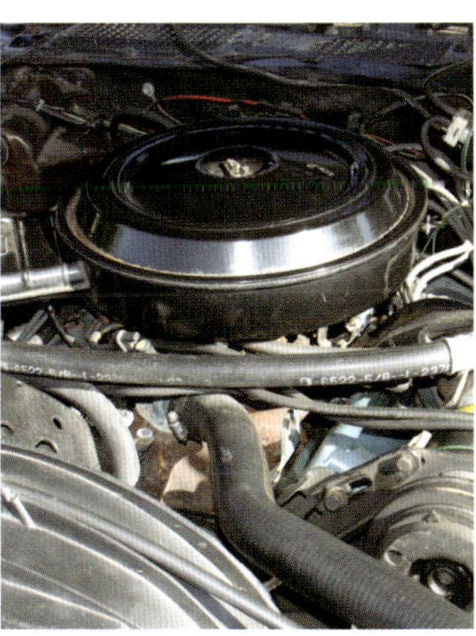

Camaro Berlinetta shown in dark blue metallic with available aluminum wheels and white lettered tires which are supplied by various manufacturers.

Camaro Berlinetta

We've brought into this world a Camaro that has layer upon layer of sumptuous comfort and ease.

Berlinetta. It's the bold signature of an extraordinary new Camaro.

First and foremost, Berlinetta has the same engineering features most other Camaros are born with. Road-hugging, finely tuned suspension. An aggressive 4.1 Litre 1-Bbl. L6 (250 Cu. In.) engine. The streamlined body that parts the air with nimble ease.

But beyond that, Berlinetta has been given a whole new level of good looks and riding comfort. A deluxe insulating package helps keep out wind and road noise. Engine mounts have been custom tuned. The shocks revalved. Front spring isolators and large body mounts smooth rough roads and help provide a lush, balanced ride.

Plus, you'll find newly designed front bucket seats, with custom-tailored fabrics and cushioned comfort that seems fathoms deep. There's a bright-accented Special Instrumentation panel, too, with tachometer, voltmeter, temperature gage and electric clock.

Outside, there's a bright new Berlinetta grille to let the world know that something remarkable is coming on through. Plus dual pin stripes, color-keyed custom styled wheels, white stripe tires, dramatic black-painted rocker panels and new Berlinetta identification on front end, trunk lid and rear roof panels.

1979

Berlinetta Coupe

The new Berlinetta replaced the Type LT, slotting in as Camaro's stab at cool elegance. Available engines ranged from the 250-cid inline 6 to the 350 V-8. Because Berlinetta didn't emphasize performance (no sport suspension was offered with it), the car carried cheerful brightwork at the upper and lower grille and on the moldings around windshield and backlight. Standard, color-keyed steel wheels could be replaced with aluminum for an additional $172. (The aluminum wheels cost $342 with the Z28.) Dual horns and a vinyl interior, options on other Camaros, were standard with Berlinetta. This example, with Light Green body and Medium Green interior, came off the line with the 130-hp 305 V-8, plus optional Soft Ray tinted glass and knit-cloth seat inserts.

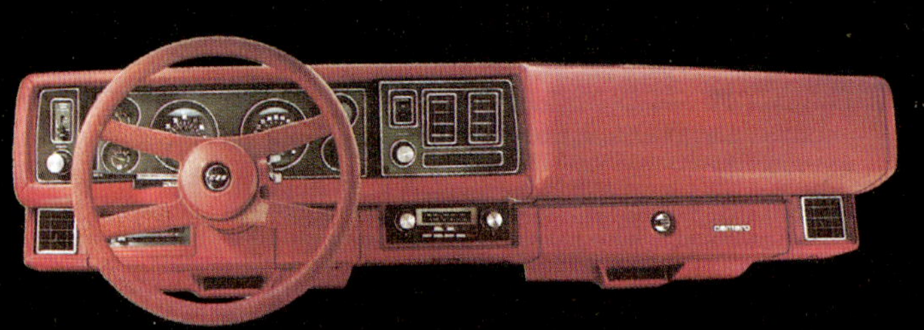

To make up the $431 price difference between Toyota Celica GT and Camaro, you'd have to go 143,800 miles.*

Camaro. Beneath the sport, more sport.

Underneath Camaro's eye-catching looks is a proven balanced suspension system with a front stabilizer bar to help control body lean during sharp turns, multi-leaf rear springs and steel-belted radials.

An efficient and economical 3.8 Liter engine is rated at 20 EPA estimated MPG (see page 11). Dualjet carburetion helps provide impressive power and quick, even response. Other standards include self-adjusting front disc/rear drum brakes and power steering. Inside, contoured full foam front bucket seats and bucket-style full foam rear seats. Plus thick, cut-pile carpeting throughout.

Four ways to go.

Your choice of four ways to get Camaro road magic. Sport Coupe, the most affordable Camaro of them all. The Rally Sport with its special exterior paint pattern. Berlinetta, for sport with a luxury accent. And the ultimate Camaro, the Z28.

Camaro Z28

Camaro Sport Coupe

*Manufacturer's Suggested Retail Price/Port of Entry price: Camaro Sport Coupe (as shown) $5,998. Toyota Celica GT Coupe $6,429. EPA est. MPG: Camaro 20, Toyota 21. For details and other comparisons, see page 11.

7

1980

Z28 Coupe

Rather than sell a detuned version of a single available Z28 engine in California, Chevy elected to give Camaro two distinct Z28 motors for 1980: the familiar 350-cid V-8 (now rated at 190 hp) for 49 states, and the 165-horse 305-cid eight for the Golden State (where the only allowable transmission was the TH350 automatic). Z28 graphics became vivid this year, with fluorescent, three-color side stripes that narrowed as they ran forward onto the front air dam. The stripes were widest at the aft end of the doors, where "Z28" was prominently displayed. A Bright Yellow body was particularly flamboyant. A functional, rear-facing hood scoop used solenoids to activate the intake door, but motive power was merely okay: 60 mph came up from rest in 8.5 seconds.

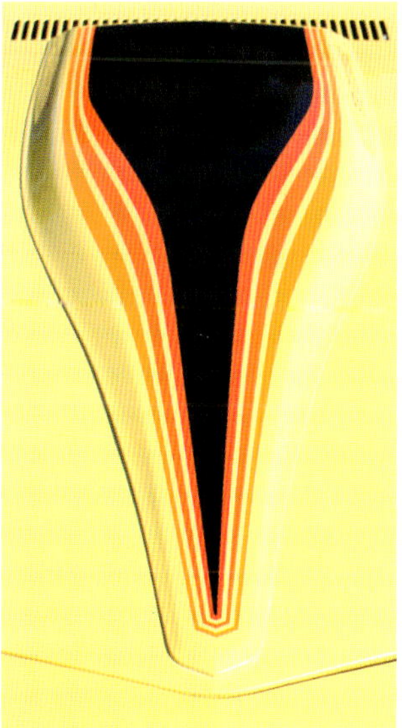

1981

Yenko Turbo Z

At the end of the Seventies, Don Yenko and his product manager, Ken Mason, decided to make a super-Camaro using Chevy's 350 V-8 and other existing components—and retain the original factory warranty, too. Turbo International, a Garden Grove, California, company, provided a water-injected turbo unit that boosted compression in Chevy's stock (but recalibrated) Quadrajet four-barrel carburetor. Turbo lag was virtually nonexistent. An automatic, which Yenko felt was better suited to the turbo than a manual, was the only available transmission. A stock '81 Z28 trundled through the quarter in 17.1 seconds at 79 mph; the Yenko Z pulled 14.52 seconds at 97.81 mph—impressive for a heavy car of that era. Two Yenko flavors were offered: Stage 1 ($11,300) and Stage 2 ($17,500), the latter with heavy stabilizer bars and other suspension upgrades.

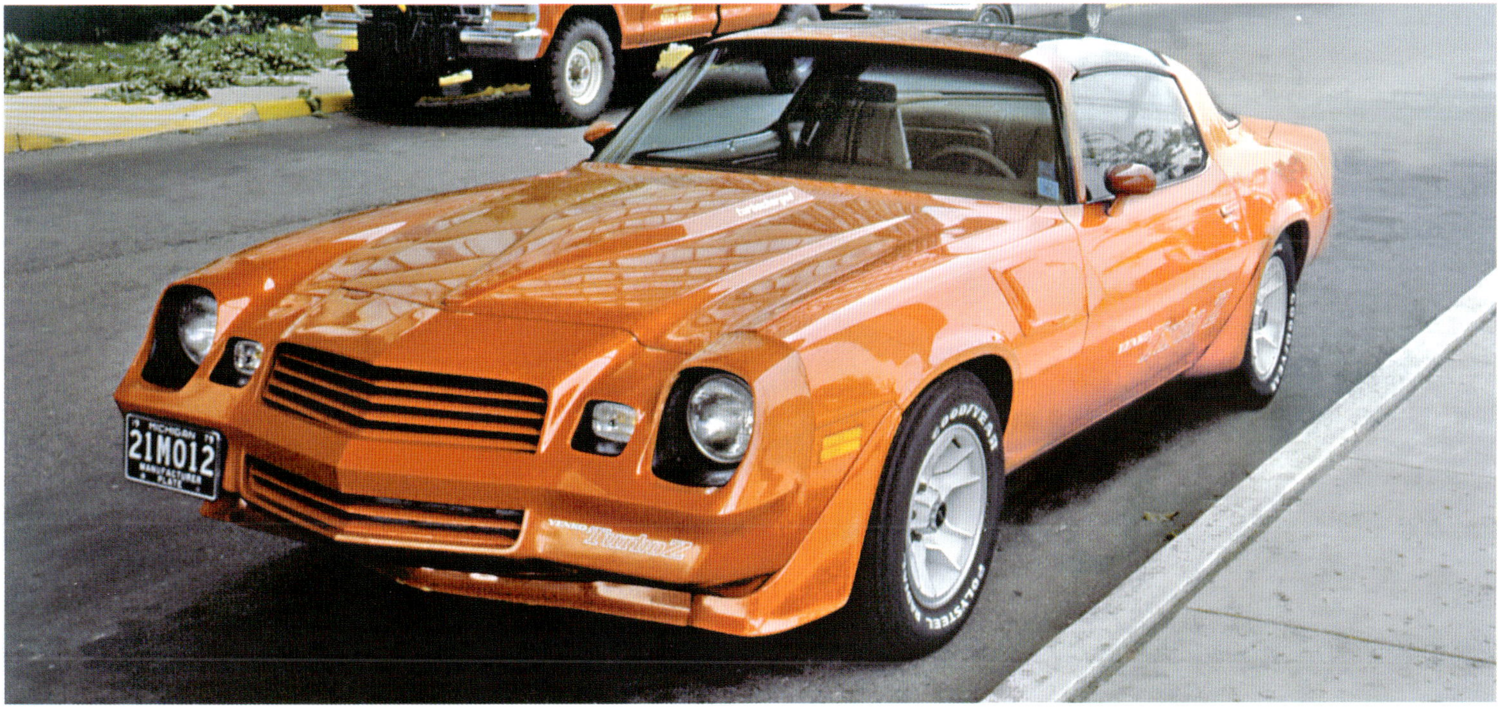

YENKO
Turbo Z

Turbo Z
21M012

1981

Berlinetta Coupe

Because the Rally Sport was dropped for '81, only the Z28 and Sport Coupe remained as stablemates of the Berlinetta. At a time when the Z28 pursued an exceedingly aggressive appearance, the restrained Berlinetta was an appealing alternative. Standard transmission for Berlinetta and SC alike was a 3-speed manual (the final time that trans would be offered). For an additional $133, a more flexible 4-speed unit that got considerably more out of the standard 110-horsepower 229-cid V-6 was available. That mattered because a V-6 Berlinetta weighed 3275 pounds—only 51 pounds less than an '81 Impala V-6 coupe. Berlinetta's wire wheel covers were standard by this time and could be protected with optional locks. This Dark Brown example has the 267-cid V-8, rated at 115 hp.

USA1

COMMEMORATIVE EDITION PACE CAR.
CAMARO
EVEN AT INDY, IT'LL HAVE THE COMPETITION CHASING SHADOWS.
This year, you'll see 33 of the finest drivers in the world chasing the shadow of the Z28 Camaro when it reappears as the Indy 500 Pace Car.
Now you can own a street version of this special Z28 Camaro in an edition so limited that only one will be offered to each American Chevrolet dealership.
The Commemorative Edition Camaro includes all the technologically advanced engineering features that made the Z28 Camaro *Motor Trend* magazine's "Car of the Year."
• Ground Effects technology, similar in theory to current Formula One practice, which literally pulls the car to the surface in some driving and road conditions.
• A 62-degree windshield rake, which helps the Z28 achieve a .369 coefficient of drag, even with downforce-inducing air dam and spoiler.
• Totally redesigned suspension, front (MacPherson-strut-based) and rear (coils, trailing links, torque tube and Panhard rod), that enabled the Z28 to generate .8g-range cornering power on GM skid pads.
• A healthy 305 CID version of the smallblock V8 can be equipped with optional new dual Cross-Fire fuel injection. (Not available in Calif.)
The Commemorative Edition Pace Car Z28 Camaro. At Indy, it'll have the competition chasing shadows.
Some Chevrolets are equipped with engines produced by other GM divisions, subsidiaries, or affiliated companies worldwide. See your dealer for details.
GM
INDIANAPOLIS
500
Z28
Chevrolet

1982

Indy Pace Car Replica

Model-year 1982 ushered in Camaro's third design generation. Because of evolving industry trends and buyer preferences, a front-drive Camaro was considered early in the development process, but Chevy elected to stay with the familiar rear-drive performance car setup. Chevy engineers and designers shaved seven inches from the wheelbase and reduced the car's weight by nearly 500 pounds. Quad headlamps were one highlight of the new, chiseled body. For the first time, a Camaro model—the Sport Coupe—came standard with a four-cylinder motor, Chevy's venerable inline "Iron Duke." Camaro paced the Indy 500 for the third time, and Chevy built 6360 Indy Pace Car replicas, based on the real deal seen here. The replicas could get either of the 305-cid V-8s available with the Z28, rated at 150 and 165 hp.

MORE THUNDER FROM THE CHEVROLET THAT TOOK THE PERFORMANCE WORLD BY STORM.

GM

INTRODUCING THE NEW 5-SPEED, 5-LITER Z28.

Camaro Z28 first became *Motor Trend's* 1982 "Car of the Year." Then, the Indy 500 Pace Car.

And now, the competition is in for some more rough weather. With the new 5-speed, 5-liter Camaro Z28.

More power and more gears than last year. To deliver lightning-like response.

Plus, all of the advanced engineering features that helped make Camaro the hottest-selling 2 + 2 sport coupe on the road today.* Including road-hugging ground effects technology and a shape so sleek, only its shadow can match it.

The new 5-speed, 5-liter Camaro Z28.

It's going to be nearly impossible for the competition to steal any of its thunder.

Some Chevrolets are equipped with engines produced by other GM divisions, subsidiaries, or affiliated companies worldwide. See your dealer for details.

*Source: Ward's Automotive Reports, Specialty Subcompact Segment, October, 1982.

Let's get it together... buckle up.

USA-1 TAKING CHARGE

CAMARO · CELEBRITY · CAVALIER · CHEVETTE · CITATION · MALIBU · MONTE CARLO · CAPRICE · CORVETTE

1983

Z28 Coupe

The big Z28 news for 1983 was a new, uprated V-8 that came on line in April of the calendar year, a 305-cid HO (high output) unit producing 190 horsepower—a 15-hp boost over the previous top 305. Fuel fed through a Rochester four-barrel carburetor, a retro technology that nevertheless produced more power than Chevy's sophisticated Cross-Fire Injection system. A cast-aluminum intake manifold was a high-tech touch. A dual-snorkel air cleaner was linked to air inlets in the headlight openings. Because of the engine's late introduction, not all 1983 Camaro literature mentions it—but the HO was very real, finding more than 3200 buyers in 1983. Manual transmission was mandatory with this motor. Z28 would do away with fiberglass hoods after this year, and no Camaro model would offer one until the SS, in 1996.

CAMARO

CAMARO

CONSUMER
GUIDE

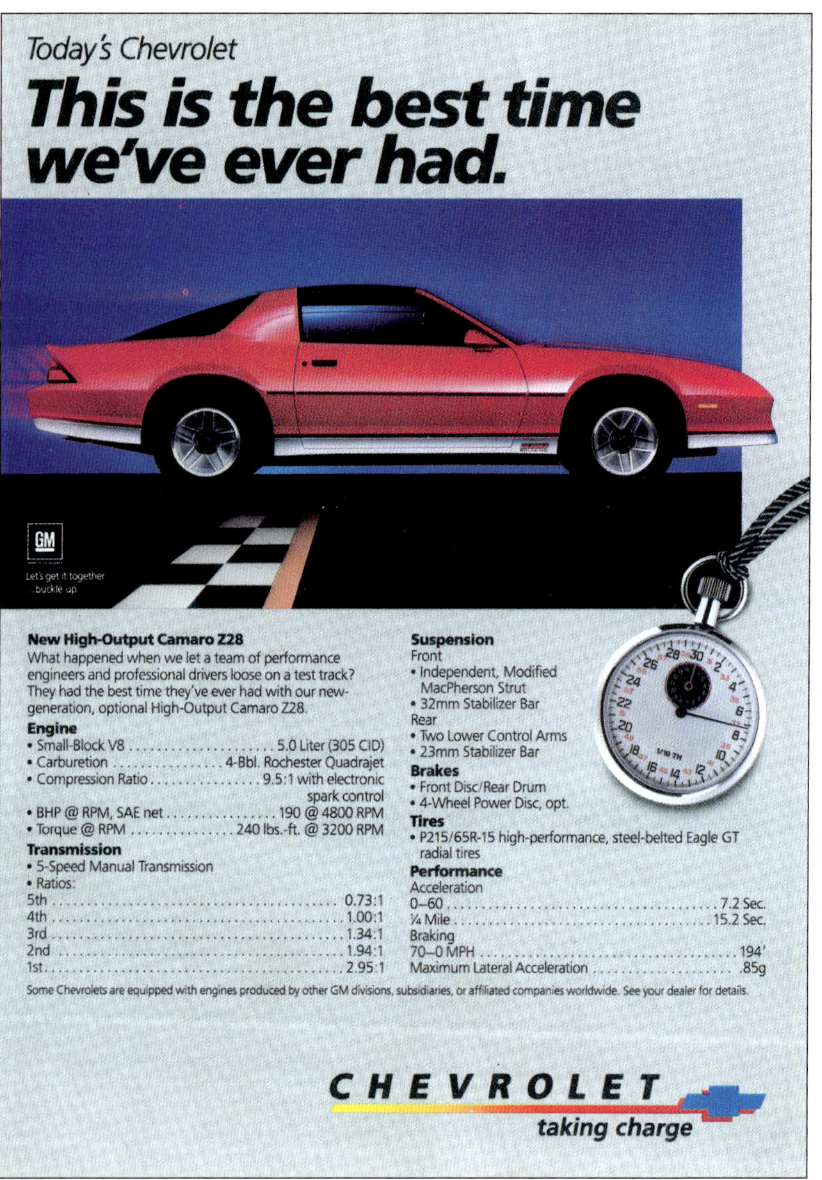

1984

Z28

The third-generation Z enjoyed considerable success in the marketplace, and in the buff books, too: *Car and Driver* magazine selected the '84 Z28 as America's best-handling car, edging out the Corvette on the strength of superior all-around road manners. The Z28's ride was subtly damped this year—a move that apparently did nothing to diminish the car's on-road stickiness. The very pleasing carbureted 305 V-8, rated at 190 horsepower, continued as the top Z28 power-plant (a 150-hp variant was standard), and available automatic transmission came on line this year. Other changes to the Z28 were minor: a visor mirror and tri-color rocker panel moldings in charcoal, silver, and red. Among this year's Z28 options were a performance axle, electric mirrors, and a $507 stereo with AM-FM, cassette, power booster, noise reduction, and a five-band equalizer.

QP 107

QP 107

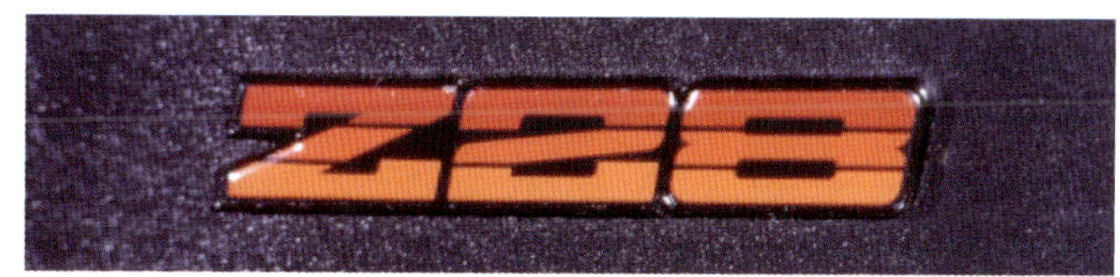
Z28

YOUR WORST NIGHTMARE.

FEEL THE POWER.
Standard 5.0 Liter 4-Bbl. V8 and 5-speed.

Optional 5.0 Liter V8 with Tuned-Port Fuel Injection. Available for the first time, with a 5-speed manual transmission. Zero to 60 in 7.0 seconds.* A 4-speed automatic overdrive is optional.

EXPERIENCE THE HANDLING.
A modified MacPherson-strut front suspension provides increased caster for precise steering. A beefy front stabilizer bar is standard.

At the rear, a torque-arm suspension with Delco/Bilstein gas-pressure shocks and available 4-wheel disc brakes.

Standard P245/VR50-16 Goodyear Eagle GT unidirectional high-performance tires on 16" aluminum wheels help IROC Z28 achieve .88g of lateral acceleration on the skidpad.*

TAKE CONTROL.
Special instrumentation with tachometer, temperature gage, oil pressure gage, voltmeter and trip odometer.

Thick leather-wrapped steering wheel and shift lever await your commands.

New optional leather seating surfaces give your IROC-Z that glove-soft feel of luxury.

Optional high-output Delco-Bose stereo system delivers superb high-performance sound.

*Performance figures compiled by a professional driver on a GM test track.

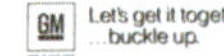

CAMARO IROC-Z

1986

IROC-Z

Grilles, front fascias, tail lamps, and air dams that had been freshened for 1985 carried on for Camaro in '86. The IROC Sport Equipment Package, named for the International Race of Champions—a body that campaigned Camaros with official Chevy sponsorship—made its debut for 1985 as an add-on package available with the Z28. For 1986, the IROC goodies added $659 to the $12,316 price of a Z28; buyers drove away with a lowered body, beefed-up suspension front and rear, reinforced front frame rail, and IROC graphics. Exclusive to the IROC was a 190-hp version of the 305 V-8, with manual transmission only. The backlight louvers on this car are factory equipment, and not an aftermarket item.

1986

IROC-Z T-top

A 215-horse, 275-pound/foot version of Camaro's 305 V-8 became an available IROC option at midyear—with automatic transmission only. The 190-horsepower version could sprint from zero to 60 in about 7.0 seconds; the 215-hp motor's performance was comparable. About 50 IROC Zs fitted with Chevy's 350-cid, 220-hp V-8 (L98) were sent to selected dealers. By now, the federal government mandated a third brake light on every American car; Camaro's (called a central high-mount stop lamp, or CHMSL) was located at the top of the rear glass, near the roofline. Although most '86 Camaros had the dreary 85-mph speedometer, a small number were fitted with a 140-mph gauge. The removable roof panels that created the so-called T-top were an $846 option.

IROC-Z

675 EPD

IROC-Z

Z28

The Camaro philosophy proudly holds up Z28 as its definitive statement--Z28 is everything Camaro is meant to be, only more. Faster, bolder, sexier.

Z28 PERFORMANCE.

- *Standard 5.0 Liter V8 with 4-barrel carburetor.*
- *Optional 5.0 Liter V8 with Tuned-Port Fuel Injection.*
- *Standard 5-speed manual transmission.*
- *Optional 4-speed automatic transmission w/overdrive.*

Z28 HANDLING.

- *Z28's acclaimed Sport suspension blends compliance with sports-car control. Componentry includes larger diameter front and rear stabilizer bars, specific spring rates and specific steering gear ratio.*
- *P215/65R-15 Eagle GT tires on 5-spoke 15" x 7" aluminum wheels are standard.*

Z28 STYLE.

- *Distinctive front air dam and "ground-effects" rocker molding design.*
- *Specific multi-color wide accent lower body striping.*
- *Rear hatch-lid spoiler with integral high-mounted stop lamp.*
- *Bold front facia, black grille and simulated hood louvers.*

NEW Z28 OPTIONS.

- *Glove-soft leather seating surfaces help make this the most luxurious Z28 ever. (Available with optional Custom Interior only.)*
- *Opt for the high-output Delco-Bose Music System and enjoy sound reproduction you've got to hear to believe. Drive the 1987 Z28 and you'll realize that Camaro can be summarized in two words: exuberant fun.*

Camaro Z28 in Yellow.

Camaro Z28 instrument panel.

1987

Z28 Coupe

Camaro juggled its lineup for 1987, dropping Berlinetta and replacing it with a sportier new model, the LT. The Z28 returned with a 165-horse 305 V-8 as its base engine. The previous uplevel 190-horse 305—a high-output, carbureted motor—was gone, and replaced with a port-fuel-injected 305 developing 215 hp. That engine could be had with a four-speed automatic transmission, or with the previous five-speed manual. The fuel-injected V-8 was surprisingly quick given its relatively modest 215 horsepower: 6.6 seconds zero to 60 mph, and 14.9 seconds through the quarter, at 95.0 mph. Only the IROC-Z got the top V-8 (350-cid, 225 horsepower), so it was clear to all that time was running out for the Z28. The badge went missing for '88 and wouldn't return until model-year 1991.

CAMARO
GHN 243
IROC-Z

1987

Z28 Convertible

Model-year '87 was the first time Camaro had offered a convertible since 1969—happy news for fans of open-air performance. By this time, Camaro convertibles were no longer built alongside coupes but were RPO T-top coupes later converted by ASC Inc. Because the ASC cars left the Chevy assembly line as coupes, confusion arose about '87 ragtops with VINs that were coded for a coupe. In its last model year—for a while—the Z28 remained a stand-alone ($13,233) that could be combined with the IROC option package to create an IROC Z28. The standard Z28 motor was a 305-cid unit, rated at 150 or 170 horsepower. A similar 305 that came standard with the IROC Z28 produced 215 hp. The Z28 came with vinyl seats, standard. An additional $28 bought cloth, and high rollers could peel off $750 for leather.

Camaro IROC-Z in Black with optional T-top roof.
CAMARO
IROC-Z
IROC-Z is the ultimate Camaro. This critically acclaimed performance coupe delivers superb response, aggressive beauty and uncompromised driving excitement.
IROC-Z POWER.
■ Standard 5.0 Liter V8 with 4-barrel carburetor. Horsepower is 170 at 4400 RPM; torque is 250 ft.-lbs. at 2800 RPM. A 5-speed manual transmission is standard, 4-speed automatic overdrive, optional.
■ Optional 5.0 Liter V8 with Tuned-Port Fuel Injection. For the first time, this powerhouse is available with a 5-speed manual transmission. So equipped, horsepower is 215 at 4400 RPM. Horsepower is 190 at 4000 RPM when equipped with the optional 4-speed automatic overdrive.
■ Optional 5.7 Liter V8 with Tuned-Port Fuel Injection. 4-speed automatic transmission w/overdrive only. Limited availability. See your dealer for details.
IROC-Z HANDLING. IROC Z28's specific tuned suspension provides exceptional cornering power and response. Here's how:
■ A modified MacPherson-strut front suspension provides increased caster for precise directional control. A beefy front stabilizer is standard.
■ At the rear, a torque-arm suspension with Delco/Bilstein gas-pressure shocks and stabilizer bar helps put IROC Z28 power to the pavement.
■ Standard P245/50VR-16 Goodyear Eagle GT unidirectional high-performance tires on 16" aluminum wheels help IROC Z28 achieve impressive .88g lateral acceleration on the skidpad.*
IROC-Z STYLE.
■ "Ground-effects" body panels and rear spoiler.
■ Specific striping and color-coordinated IROC-Z door decals.
■ Halogen fog lamps.
■ Stowaway spare tire.
■ Sport mirrors.
■ Rear spoiler with integral high-mounted stop lamp.
IROC-Z CONTROLS.
■ Special instrumentation with tachometer, temperature gage, oil pressure gage, voltmeter and trip odometer.
■ Thick leather-wrapped steering wheel and shift lever await your commands.
■ New optional leather seating surfaces give your IROC-Z that glove-soft feel of luxury. (Available with optional Custom Interior only.)
■ An AM radio with digital clock is standard. (May be deleted for credit.)
■ Select the optional high-output Delco-Bose stereo system for more superb high-performance sound.
IROC Z28 presents a truly unique opportunity for the dedicated performance enthusiast.
*On the GM test track by professional drivers.
5

1987

IROC-Z

With the Z28 on its way out, the top-dog Camaro for 1987 was the IROC-Z, still a Z28 add-on. Because of the co-op deal with the International Race of Champions, Chevy was eager for the IROC be the performance Camaro. To that end, the standard engine was Corvette's 350-cid V-8, but with cast-iron heads rather than aluminum, and exhaust manifolds of iron rather than stainless steel. Added cost: $1045. To further protect the Corvette brand mystique, the 350 was downrated by 15 horsepower for IROC installation, from 240 to 225. The 1987 IROC-Z could be optioned with a performance rear axle, oil cooler, and 4-wheel disc brakes. P245/50VR16 blackwall tires were a no-charge option. A stock IROC cranked from zero to 60 in 6.3 seconds and could run the quarter in about 14.5 seconds at 95 mph.

Camaro IROC-Z.
Now available with
solar heating.
Let's get it together... buckle up.
GM 6/60
QUALITY COMMITMENT PLAN
IROC-Z
Think of it as a built-in observatory. A standard tanning booth.
Or just think of Chevy Camaro's IROC-Z Convertible as the best way to top off a car that has virtually everything.
Optional 5.0 Liter V8 with Tuned-Port Fuel Injection, standard 5-speed overdrive, sport suspension with fat front and rear stabilizer bars and rear gas shocks, ground effects and rear spoiler, available Goodyear Eagle GTs and new 16-inch aluminum wheels, full cockpit instrumentation: including tachometer, temperature, oil pressure and voltmeter.
So crank up the stereo. Notify the neighborhood.
And when you've got their attention, flaunt it.
Let down the top and let it all hang out.
Chevy thunder now comes as a 2+2 coupe or 2+2 convertible.
V8 thunder: Available 5.0L TPI V8 = 220 HP.
0-60 MPH in 7.1 secs.
(Professional driver on test track.)
It'll do a sticky .86g on the skidpad.
6-year/60,000-mile powertrain warranty.
See your Chevrolet dealer for terms and conditions of this limited warranty.
THE Heartbeat OF AMERICA
TODAY'S CHEVROLET

1988

IROC-Z

Model-year 1987 was the Z28's last hurrah, so the IROC-Z was top-dog Camaro for '88. No Camaro motor had a carburetor this year; fuel injection was standard across the line. The 305-cid V-8 picked up a new camshaft that bumped horsepower to 195 with automatic transmission, and to 220 when mated to the T5 manual. The 350 (optional with the IROC only) also got a five-horse boost, to 230; the same engine developed 245 hp for Corvette. The other big news was that a Camaro convertible was offered for the first time since 1969, available in IROC or Sport Coupe trim. Because GM closed its dedicated Camaro/Firebird plant in Norwood, Ohio, after the 1987 production run, all Camaros now came out of the Van Nuys, California, facility.

Camaro IROC-Z Coupe.

Camaro RS Convertible.

IROC-Z Thunder

Camaro IROC-Z, the American performance legend, returns for 1990 with a choice of standard 5.0 Liter V8 power or the optional 5.7 Liter V8. IROC-Z also includes a Special Sport Suspension and distinctive exterior trim.

All 1990 Camaros feature PASS-Key,® the most effective theft deterrent system in Chevrolet history.

New for 1990 on all models: a driver side Supplemental Inflatable Restraint System (air bag), restyled instrumentation and **Scotchgard**™ Fabric Protector on seats and interior door panels.

Convertible Passion

Both RS and IROC-Z are available in an exhilarating convertible body style that creates the ultimate Camaro driving experience.

Camaro Convertible features include:

- Room for four.
- An enclosed, lockable trunk.
- Manually operated cloth top.
- A flush-fitting fiberglass boot for a neatly-integrated top-down appearance.

1990

IROC-Z Coupe

The standard IROC-Z engine for the 1990 coupe and convertible was the 305-cid, 210-hp V-8. Coupe buyers who wanted more giddyap could part with an additional $300 for the 350-cid V-8 producing 245 hp (not available with the ragtop). The “Z28 IROC-Z” dash logo was yellow, for this year only, and instrument lighting shifted from white to yellow. A driver-side air bag came on line, and all Camaros picked up a tilt steering column. Only IROC-Z steering wheels were leather-wrapped. Fifteen-inch aluminum wheels were standard, with 16-inchers optional for $520. This Bright Blue IROC is one of 62 1LE models built for SCCA Showroom Stock competition. These cars had the 350 V-8, Corvette brakes, aluminum driveshaft, gas tank baffle, and (on the car seen here) an automatic trans with overdrive.

CAMARO RS

LEGENDARY PERFORMANCE, SURPRISING FUEL ECONOMY

EPA est. **18** MPG City **27** MPG Hwy.

with 3.1 Liter V6/optional 4-spd. automatic transmission.

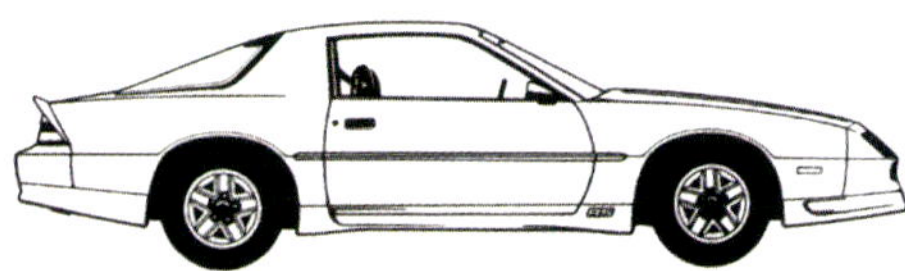

THAT'S BETTER HWY. MILEAGE THAN:
■ Dodge Daytona V6/automatic ■ Ford Probe V6 ■ Toyota Camry V6 ■ Chrysler LeBaron V6/automatic.

Other Camaro Mileage Examples:
Camaro RS V6/5-spd.: EPA est. 17 MPG City/27 MPG Hwy.
Camaro RS V8/5-spd.: EPA est. 17 MPG City/26 MPG Hwy.

MORE REASONS TO BUY CAMARO

- Choice of RS or Z28 models
- Choice of coupe or convertible body styles
- PASS-Key® theft-deterrent system
- Driver's-side air bag
- Legendary Camaro style and performance.

1991

RS V-6 Coupe

Because Chevrolet elected not to renew its contract with the International Race of Champions for 1991, the IROC-Z was discontinued. The well-loved Z28 returned to fill Camaro's performance slot. The Z and its remaining stablemate, the RS, were available in coupe and convertible forms. RS coupe production totaled nearly 80,000 this year; most, like the handsome, Arctic White daily driver pictured opposite, were powered by the second generation of GM's 60° V-6, a 191-cid, 140-horsepower unit that came standard. Intriguingly, though, a police-package RPO (B4C) could turn an RS into a faux-Z28, with the 245-hp 350 V-8 running with automatic transmission, plus limited-slip rear axle, four-wheel disc brakes, aluminum driveshaft, uprated shock absorbers, and engine oil cooler. If manual transmission was desired, the package brought the 305, rated at 230 hp.

91
CAMARO Z-28

1991

Z28 Coupe

The return of the Z28 brought a new, taller rear spoiler, nonfunctional hood bulges, and a red bow-tie grille emblem, plus, as on all '91 Camaros, freshly designed front and rear fascias and new ground-effects panels. The car perpetuated the spirit of the deleted IROC and continued to be popular with performance fans. Because the race season began in the spring, the '91 Camaro line was introduced early, in March 1990. Base engine for the Z28 was the 205-hp 305-cid V-8; an extra $300 bought the 245-horse 350, but for the Z28 coupe only (and with the RS police package; see previous spread). Z28 convertibles had the 305, mandatory.

25TH ANNIVERSARY
Camaro '70
'67 camaro
'82 CAMARO
CHEVROLET
CAMARO
1992

1992

25th Anniversary Z28

By the 1992 model year, a full quarter century had passed since the first Camaro—reason enough for Chevy to commemorate the anniversary with across-the-line dash emblems and a $175 Heritage Appearance Package (RPO Z03) that could be had with the Z28 and any other '92 Camaro. The RPO brought unique stripes on the hood and deck, a body-color grille, black headlight surrounds, and discrete decklid badge. But buyer anticipation of a redesigned '93 Camaro caused Heritage Package production to top out at a modest 8197. Production of the Z28 totaled 6451, of which 5197 were coupes. The car looked tasty in Arctic White, one of four colors exclusive to the RPO when it was ordered for installation on the Z28 (the others were Black, Purple Haze, and Bright Red).

11

TWENTY-FIVE YEARS AGO the first Camaro rolled off a Chevrolet assembly line ... and an American love affair began. Competitors have come and gone, but down on Main Street, Chevy Camaro is still USA-1.

Camaro RS for 1992 continues the tradition with enthusiasm. Consider this the practical, easy-to-own Camaro. Yet it still offers all the appeal of a rear-wheel-drive coupe that has been carefully refined through the years. A 3.1 Liter V6 is standard, and a 5.0 Liter V8 with Electronic Fuel Injection is optional on Camaro RS models.

Camaro Z28 is all action, from its 5.0 Liter V8 with Tuned-Port Fuel Injection to its Ride and Handling suspension. A domed hood and aero-formed rear wing add to the unique appeal of the Z28 Coupe. And for those who won't settle for less than the ultimate Camaro, there's the Z28 Convertible.

CAMARO MODELS

RS Coupe	Z28 Coupe
RS Convertible	Z28 Convertible

ENGINES

- 3.1 Liter V6 with Multi-Port Fuel Injection (standard on Camaro RS models).
- 5.0 Liter V8 with Electronic Fuel Injection (optional on Camaro RS models).
- 5.0 Liter V8 with Tuned-Port Fuel Injection (standard on Camaro Z28 Coupe and Convertible).
- 5.7 Liter V8 with Tuned-Port Fuel Injection (optional on Camaro Z28 Coupe).

TRANSMISSIONS

- 5-spd. manual (standard on all models).
- 4-spd. automatic with overdrive (optional on all models).

BRAKES

Power front disc/rear drum system. Four-wheel power disc brakes optional on Camaro Z28.

DIMENSIONS (in.)	Coupe	Convertible
Wheelbase	101.0	101. 0
Overall length	192.6	192. 6
Overall width	72.4	72.4
Luggage capacity (cu. ft.)	31.0	5.2
Fuel tank capacity (gal.)	15.5	15.5

Camaro RS Coupe.

1
Camaro RS Convertible. Shown with optional 16" tires and sport-styled wheels.

2
Chevy Camaro—the American performance legend—celebrates its 25th anniversary in 1992.

3
A driver-side air bag and the GM-developed PASS-Key theft-deterrent system are standard in every 1992 Camaro.

4
Camaro Z28 Coupe (foreground). Camaro Z28 Coupe with the optional Heritage Appearance Package (background).

1992

25th Anniversary RS Convertible

Of 70,007 Camaros produced for 1992, more than 63,500 were RS models. The convertible RS accounted for only 2562 units but was a handsome boulevardier that used multiport fuel injection to coax 140 horsepower from its standard 191-cid V-6. Torque was 180 pound/feet at 3200 rpm. All '92 Camaro convertibles picked up courtesy lights and three-point seatbelts for backseat passengers. This Bright Red example has the Heritage Package, which included hood and decklid stripes, and blackout headlight coves. Power front disc brakes, tilt steering, manual air conditioning, and power door locks were some of the model year's most popular options.

25TH ANNIVERSARY

29 M774

Z28.
America's Next
Classic.
Z28
Camaro Z28 Coupe in Arctic White
with optional removable roof panels.
35

1993

Z28 Coupe

Camaro kicked off its fourth generation with a car that was newly designed inside and out. Taking some of its appearance cues from Chevy's California Camaro show car, it was swoopier and more feminized in body shape than its predecessor. The cabin was more rounded, as well, and had less of a "cave" aspect than before. Camaro was undeniably new but retained such popular cues as tunneled headlights, wide tail lamps, and a dramatically vee'd nose. The body/chassis unit had steel-reinforced composite panels over a steel frame. Despite this, and some newly available options, the new Camaro weighed only about 150 pounds more than the '92. The Z28's 350-cid V-8 picked up 30 more horsepower, running now with 275. A new 6-speed manual (borrowed from Corvette) was the base Z28 trans; an automatic was optional. With the manual, 60 mph came up from rest in a hair over 5.0 seconds.

THE NEW CAMARO:
BACK HOME AGAIN
IN INDIANA.
When the 1993 Camaro Z28 takes its place at the head of the field to pace the 77th running of the Indy 500, it will be a beautiful moment in Indianapolis Motor Speedway history — and a proud moment for Chevrolet.
FOREGROUND:
1993 Camaro Z28 Coupe
with optional Indy Pace Car Package.
BACKGROUND:
1967 Camaro SS396 Indy Convertible (left).
1969 Camaro SS396 Indy Convertible (center).
1982 Camaro Z28 Indy Coupe (right).
CAMARO
OFFICIAL PACE CAR
Z28
USA1
USA1

1993

Z28 Indy Pace Car Replica

In 1993, a Camaro paced the Indianapolis 500 for the fourth time, and to celebrate, Chevy offered a $995 pace car-replica package with Z28 underpinnings. The '93 was the first Chevy pace car to go to the track straight from the factory, without modifications. To create a replica, then, was an easy matter, and pretty much cosmetic at that, with black-over-white body, painted white wheels, Indy badging and other exterior graphics, and unique seats. The Z28 engine was the 350-cid (5.7-liter) V-8, adapted from Corvette's LT1 and rated at 275 horsepower. A 6-speed manual was standard, with a 4-speed automatic on the options list. As on all other '93 Camaros, the pace car replicas had driver and passenger air bags. Production totaled 633.

1994 CHEVROLET CAMARO CONVERTIBLE

1994

Z28

The Z28 could be had as a coupe or convertible (with full headliner) this year, and came standard with a 6-speed manual mated to a 350 V-8 rated at 275 horsepower. (A variant of that motor installed in the '94 Impala SS and Caprice produced 260 hp.) Peak torque of 325 pound/feet developed at just 2000 rpm, making this Z a serious stoplight racer. The Z28 coupe, which saw production of 36,008 for '94, retailed for $16,999—which is less than $17,000, by the way. Z28 convertibles, with production of 4942, sold for $22,075. Base tires for the Z28 were P235/55R16 blackwalls. A $225 upgrade brought P245/50ZR16 rubber, also in black. A rigid chassis eliminated body flex on Z28 coupes and dramatically limited it on ragtops. The solid rear axle remained, and although the unit was far from state of the art, it performed adequately and kept retail prices in line with what Camaro fans were willing to pay.

1995 Camaro Z28

1995

Z28 Convertible

Convertibles returned to the Camaro lineup for 1994, in base V-6 and Z28 form. Not surprisingly, the ragtop Z was the most-expensive '95 Camaro, coming in at $23,095 (a Z28 coupe was $17,915), with production of 8024 units. Traction control became available on Z28 coupes and convertibles for '95; the system worked particularly well with newly optional all-weather, Z-rated tires. A tamer push rod action made the 350-cid LT1 engine considerably more relaxed at idle. In a nod to fuel economy, the Z28's standard 6-speed manual gained Computer-Aided Gear Selection that forced a shift from first to fourth in tame acceleration. But when asked to perform, the Z28 convertible handled the zero-to-60 sprint in brisk fashion: 6.2 seconds, according to *Motor Trend* testers.

Camaro. Are You Ready To Rock 'n Roll?

You know what this Chevy is all about. Unleash it, and you'll remember why you wanted that license so badly.

Yeah, Camaro's got an attitude, and the 1997 cranks up the volume some more. New 30th Anniversary edition. New interiors. New colors. This is, without question, the bad boy of sport coupes.

Your choices include Camaro or Camaro RS (powered by the 200-hp, 3800 V6) or the legendary Camaro Z28 (packing the 285-hp, LT1 V8). All are available in Coupe, T-Top Coupe or Convertible versions.

Rock 'n roll. Fast food. Fast cars. It's an American thing. It's why we love Camaro.

Camaro Coupe with optional aluminum wheels. Standard features include a 3800 V6 engine and 5-speed manual transmission.

Camaro Z28 Convertible with the 30th Anniversary Package and other options.

1997

30th Anniversary Z28 Coupe

Arctic White with Hugger Orange stripes—that was the exterior color combo that came with RPO Z4C, Camaro's $575 30th Anniversary package, exclusive to the Z28, running this year with 285 horses churned out by the 350 V-8. Purchasers of the anniversary RPO had their choice of white leather seats, or white leather with houndstooth inserts (the latter a nod to the '69 Camaro Indy pace car). All '97 Camaros had stitched or stamped 30th Anniversary logos on the headrests, as well as new tri-color taillights, a redesigned instrument panel, and enough cup holders to keep everybody hydrated for miles and miles. Delco/Bose stereos went away and were replaced by a solo Delco system cunningly noted on the options list as "Audyssey."

You could buy a more practical car,
but then you'd have to drive it.
New Camaro Z28. Genuine Chevrolet
Okay, we admit, with 305 hp, a six-speed transmission* and the top down, it's anything but practical. But can you think of a better way to get your hair all messed up? Well... besides that.
Call 1-800-950-2438 or visit www.chevrolet.com *No-cost option. Camaro Z28 is a registered trademark of the GM Corp. ©1997 GM Corp. Buckle up, America!

1998

Coupe

Midway through the fourth generation, Camaro was given a reskin that brought a completely redesigned hood, front fascia, grille, and headlights. This was the first "open-mouth" grille since 1989 and recalled the grillework of Camaros from the early Seventies. The RS was dropped, though its essentials could still be had (except with SS models) in a new appearance package coded Y3F. The option brought an extended front fascia, a rear-spoiler extension, rocker and rear fascia moldings, and 55R16 tires on 160-inch aluminum wheels. Bosch four-wheel disc brakes were now standard across the Camaro line. Despite the re-do, and 305-horse V-8s (adapted from Corvette's aluminum-block LS1 unit) for Z28 coupes and ragtops, total Camaro production was 54,026—down from 1997 by about 6000 units.

CAMARO
THIRTY-FIVE YEARS OF HONEST AMERICAN PERFORMANCE.
Through changing times and changing rules, Camaro has never compromised. To everyone who has ever owned — or ever dreamed of owning — a Chevy Camaro, we dedicate the limited-edition 35th Anniversary Camaro. It's a very special way to celebrate 35 years of honest American performance.
2002 Camaro Highlights:
35th ANNIVERSARY SS CHEVROLET CAMARO
35th Anniversary Camaro SS interior
FOR MORE INFORMATION
visit our web site at chevy.com
9
35th Anniversary Edition Camaro SS in Bright Rally Red

2002

35th Anniversary SS

Production of the 2001 Camaro, at just 29,009 cars, was the lowest in Camaro history, so Chevy's 35th birthday brought some uncertainty. To commemorate nevertheless, Chevy offered an SS-based 35th Anniversary model with a T-top (if a coupe) and a black soft top if an open Camaro. The Anniversary cars were Bright Red with black leather interiors; headrests had anniversary embroidery. To suggest Camaro's competition heritage, exclusive dual silver stripes culminated at rear deck and hood in a checkered-flag design. Unique badging on the front fenders, and at nose and tail, further identified the package. Total Camaro production for the '02 model year, at 42,098, was a marked improvement over '01 and brought the numbers in line with figures for 1999–2000. And yet, Camaro production was suspended after the '02 model year.

You can tell a lot about a man
by the cut of his sheet metal.
Price as shown, $26,895.[2] 1 Based on EPA estimate. 2 MSRP. Tax, title, license, dealer fees, and optional equipment extra.
Camaro is a registered trademark and Chevy is a trademark of the GM Corp. ©2009 GM Corp. Buckle up, America!
304 HP V6 ▸ 29 MPG HWY[1] ▸ STARTING AT $23,040[2] ▸ CHEVY.COM/CAMARO ▸ ALL NEW CAMARO

2010

Coupe

Development of a reborn Camaro was undertaken at the beginning of 2005 and accelerated dramatically in midyear, when executives asked for a running prototype by October. The visual baseline for the new design was the 1969 Camaro, with some underpinnings borrowed from the Pontiac G8 sedan. Top engine, which was exclusive to Camaros in SS trim, was a 426-horse 6.2-liter V-8 mated to a 6-speed manual transmission. Thus equipped, a Camaro rocketed from zero to 60 in 5.0 seconds. The same motor, detuned to 400 hp and linked to an automatic transmission and GM's Active Fuel Management cylinder deactivation system, was nearly as quick. A 3.6-liter, 304-hp V-6 was also available, and ran 0 to 60 in a sterling 6.0 seconds. Power, retro styling, and sophisticated engineering made the '10 Camaro a critical and sales smash.

CAMARO

2011

Convertible

Although supplier issues delayed the debut of the new Camaro convertible until the second quarter of 2011, public enthusiasm remained high, particularly when "buff" magazines published spy photos of prototypes during the summer of '10. A formal reveal was held in November 2010, at the Los Angeles Auto Show. Prices that were announced early in 2011 ranged from $29,150 (219.7-cid, 312-hp V 6) to $39,650 (378-cid, 426-hp V-8). When buttoned up, then, the 2011 Camaro ragtop has a narrow greenhouse offering the same over-the-shoulder visibility issues as the coupe. A *Motor Trend* test of an early example got a zero-to-60 time of 4.9 seconds.

Camaro